DEVOURED

DEVOURED

THE EXTRAORDINARY STORY OF KUDZU THE VINE THAT ATE THE SOUTH

Ayurella Horn-Muller

LOUISIANA STATE UNIVERSITY PRESS / BATON ROUGE

Published with the assistance of the V. Ray Cardozier Fund

Published by Louisiana State University Press
lsupress.org

Manufactured in the United States of America
First printing

DESIGNER: Michelle A. Neustrom
TYPEFACE: Calluna
PRINTER AND BINDER: Sheridan Books, Inc.

Jacket illustration created by Michelle A. Neustrom.

Excerpt from "The Terror of Kudzu," by Edward Francisco, is used by permission
of Edward Francisco and *Appalachia Bare.* Copyright © 2019 by *Appalachia Bare.*

Excerpt from "Kudzu," by Saeed Jones, from *Prelude to Bruise*
(Coffee House Press, 2014), is used by permission of Saeed Jones and
Coffee House Press. Copyright © 2014 by Saeed Jones.

Cataloging-in-Publication Data are available from the Library of Congress.

ISBN 978-0-8071-8200-0 (cloth: alk. paper) — ISBN 978-0-8071-8240-6 (pdf) —
ISBN 978-0-8071-8239-0 (epub)

For Alex—who introduced me to the untold magic,
mayhem, and mystery behind one bristly vine.

On closer look
it becomes what we most
despise.

—EDWARD FRANCISCO,
"The Terror of Kudzu"

I won't be forgiven
for what I've made
of myself . . .
If I ever strangled sparrows, it
was only because I dreamed
of better songs.

—SAEED JONES,
"Kudzu," in *Prelude to Bruise*

Contents

Photographs follow page 84.

DEVOURED

INTRODUCTION

Lush fields unfold under a midday sun, emerald hues glinting off blades of bermudagrass. A swarm of dragonflies ascends into unhurried clouds, indiscernible wings moving back and forth in a mechanical dance, armed with needle-like precision. I step forward with false bravado, watching a damp carpet of soil cushion my foot, with a heightening degree of trepidation. In and out my breath stutters as I force myself to move, my tunnel-vision sight trained on a thicket of trees a couple hundred yards ahead. Old bottles and faded pieces of trash litter the ground I walk upon, evidence of the humans who continue to invade this ecosystem flashing in my peripheral view. My destination is a plot of oaks, an untamed haven bordering a nature reserve and a bustling freeway, just outside of Athens, Georgia. Woven within the wild expanse of greenery is a timeless relic, a bewildering being enshrouded in over a century of magic and mystery.

Here will mark my first up-close encounter with kudzu—a bristly, leafy vine famed in the American South for its propensity to smother other plants at an unparalleled pace. Its lasting influence on popular culture can be traced back to the twentieth century. Ever since, the "vine that ate the South" has been immortalized in poems, books, movies, and even nursery rhymes once hysterically giggled by generations of American children playing in fields overrun by the very treacherous plant they were warned about.

I have no way of knowing if the stories about kudzu's thirst for destruction and unstoppable growth are spun from threads of truth. What I understand about the tickings of the terrestrial biome I find myself in is limited to the musings and recollections of those who came before me. Firsthand accounts chronicling daring expeditions into old-growth forests and epic showdowns with grizzly bears line my bookshelves back home. A soothing monologue voices over dreamlike moving pictures capturing the perennial dance between predators and prey; images melding with underwater footage drenched in enhanced colors as courageous explorers spend a year tracking a wild octopus[1] off a South African coast or documenting the rapid decline of coral reefs[2] in the Caribbean. Set amidst green palaces and aquatic realms, these recordings of others' adventures are all I know of the wild: the nearly one-quarter of the planet's terrestrial surface[3] unharmed by human activities. Beyond those reports is nothing but my incandescent imagination, where a war wordlessly wages within. One part of me desires a stronger connection with nature—while the other ferociously fears it.

It's this precipice I stand before, my perceptions of this environment vanishing into an imagined impenetrable fog. A hologram of despair rises slowly from the forested floor as I continue roving ahead, pulse ricocheting as my phobias threaten to swallow me whole. Those days of finding comfort and stability in the sheltered, carefully structured human ecosystem I grew up in are but a memory. There, I watched the world come alive from indoors, every interaction within my control. Here, wildlife roams freely on every visible surface—and in many hidden ones. This domain I am walking into is entirely unfamiliar. Like a bottlenose dolphin that's strayed too far from its pod, I feel exposed, as if I were navigating strange waters swarming with apex predators. I have no semblance of power in this uncharted space. My dread has gripped me in a chokehold, overruling my ability to be calm and collected in this untamed thicket. Caught in a conflicting wave of apprehension and curiosity, here in Georgia's Oconee Forest Park, I am hundreds of miles from the concrete jungle I call home. All in pursuit of one impressionable vine.

Where I grew up, swaths of kudzu reigned supreme. Driving along any number of interstates in Florida meant witnessing an overabundance of the perennial weed, found everywhere from Tallahassee to Miami, draped lan-

---

1. *My Octopus Teacher* is a 2020 Netflix documentary, directed by Pippa Ehrlich and James Reed, that follows Craig Foster's year spent diving with one octopus. https://www.imdb.com/title/tt12888462/.

2. *Chasing Coral* is a 2017 documentary, directed by Jeff Orlowski-Yang, that records the disappearance of the world's coral reefs. https://www.imdb.com/title/tt6333054/.

3. This measurement excludes Antarctica. James E. M. Watson et al., "Protect the Last of the Wild," *Nature* 563 (2018): 27–30. https://doi.org/10.1038/d41586-018-07183-6.

guidly over trees dotted along roadways, open fields, and abandoned build-ings. As familiar and forgettable as a palm frond, kudzu is as synonymous with the South as Spanish moss—the only other plant commonly spotted envelop-ing everything from train tracks to powerlines. But I was never drawn to Span-ish moss—another aesthetically compelling invasive—the same way I gravi-tated toward kudzu and its towering leafy palaces. Perhaps it was personal. While I was raised to live in fear of the unmapped plots of nature's playground, what struck me about the proverbial vine was the story embedded within it. For almost a century, the plant has danced under an American limelight, as a sweeping fixture of southern culture. It wasn't long before I was hooked on the antiquated fables once whispered at twilight into the ears of sleepy children, warning of one plant's superhuman-like strength or the way it smothered any-thing unlucky enough to stray into its path of vengeance. What drew me to kudzu was the way it was humanized—how one prodigious little vine mor-phed into a daunting creature that eventually ate an entire geographic region. An influence that has withstood the test of time. From there, I fell headfirst in love, as the muse that sowed the seeds of these harrowing visions burrowed into my soul, begging to be investigated beyond these one-dimensional char-acterizations. I began to wonder: How could we ascribe such fantastical qual-ities to one weed? Could this noteworthy shift in cultural perception have something to do with the plant's distant origins? And what might this say about the American relationship with racial and ethnic diversity, with this country's condemnation of plants—and people—with roots from elsewhere?

Over the course of three years, I mined historical accounts, peer-reviewed scientific papers, and government reports on the inner workings of the vine that devoured an entire portion of America. I learned everything I could about how a plant, native in East Asia but invasive in the United States, came to be such a persistent, complicated emblem of the South. I tracked down doz-ens of experts behind modern research on kudzu to interview them about their work, asking questions to determine what we conclusively know about it—and what we don't. I spoke to people carrying conflicting views on the plant's role in the American landscape, and I learned more about their origin stories, examining why they themselves see something that everyone else is quick to disparage. Emerging from my descent into these global archives is a remarkable tale—one that recollects, reminisces, and rationalizes the ways kudzu morphed from a glorified, miraculous solution for soil erosion into a monstrous archetypal villain of the landscape: a dramatic shift that uprooted the way an entire geographic area, as well as a country, perceived a vine with a genesis in a faraway place, and how that contributed to cultural and racial dissonance for years to come.

Perhaps that dichotomy is what first truly attracted me to the narrative of a noxious weed. After all, kudzu's war with cultural perception in the States is something I identify with. A second-generation American,[4] I grew up in a household where Asian traditions mixed with Mediterranean ones, several languages were spoken freely, and diverse cultures blended, producing racially ambiguous results. When I was just a kid in a sea of youthful peers blissfully void of prejudice, being unique was a novelty. My skin, my hair, my eyes—each were distinct parts of me that were praised, and sometimes coveted. It was even deemed darling the way I sometimes confused other languages for English, a habit I had picked up unknowingly.

But with time comes wisdom. Or at least a candid awareness. The older I got, the more the reception to the things that set me apart soured, like a flame violet left to the mercy of the unruly sun. People began commenting on the distinction of my look, as assumptions were leveled at my backstory. They'd ask: "Where are you *really* from?" Strangers balked at pronouncing my name: "Wait, *how* do I say that?" People I considered allies even started pointing out the way my mom spoke: "What a *strong* accent she has!" Perhaps, in reality, nothing had changed, and all of those things had always been staples of my existence, simple byproducts of what it means to live in a nation as inherently anti-immigrant as this one, but with each year of life, I began to grasp how they ostracized me from the rest. Perhaps I just became more observant of the strands separating my peers and myself—differences that had always been there, simmering below the surface.

Something I had once been glowingly proud of morphed into a layer of me I acutely wanted to hide. By the time I started high school, I stopped inviting people over, unsure how they would react when they learned that my dad wore a sarong around the house ("Is your dad in a *skirt?!*") or how they

4. The Pew Research Center defines a second-generation American as someone "born in the United States with at least one first-generation (immigrant) parent." https://www.pewresearch .org/social-trends/2013/02/07/second-generation-americans/.

would scoff at the obtrusively multiracial dishes my parents prepared. Beyond the safety of that space, those legacies became sources of shame. I learned to avoid lingering in the treacherous daylight that would deepen my already too-brown skin tone, in the hopes that I could somehow become paler and make my body less offensive to the shades of white around me. I turned to makeup to try to disguise my slanted eyes, which others would remind me of when they nonchalantly deployed slurs to describe my race. It wasn't long before I even tried shortening my name to better fit in with the monosyllabic American girls I found everywhere I looked. It, like everything else, branded me as different. I became convinced that if I tried hard enough, everyone might believe that I was just like them. Perhaps I could make the world kinder, more accepting, once I changed my name and camouflaged the ways I stood apart.

Maybe it's the consequence of being treated like a consequence. All too often, I have longed to shed my distinct ancestry; to be worthy of an American stamp of acceptance. Too many times, I have longed to remind those around me: "I am here because I have a right to be!" So why did it feel like I was trespassing, as if I'd crossed some invisible line? It never made much sense to me how anyone from a country with an origin story like the United States could be so opposed to welcoming people with beginnings beyond these borders. Much like kudzu, the former pulsing heart of a green, southern kingdom, I felt as if I was diminished and criminalized by the very living descendants of communities that had once welcomed the arrival of those similar to me. It's hard not to see parts of myself reflected in the stand of kudzu I stand before. First embraced, then shunned and eventually accepted with lasting disregard. Unable to escape the otherness that brands us—the ones from distant places. But still thriving, against all odds, labels, and unjust expectations.

Once beloved, then feared, and eventually just tolerated, kudzu can be found nearly everywhere across the American South. Populations have long persisted in Florida, Alabama, Georgia, South Carolina, North Carolina, Tennessee, Texas, Louisiana, Kentucky, Virginia, Mississippi, and Oklahoma. Although it has been part of the natural topography in one American region known for its dense wetlands and unforgiving humidity, plots of land engulfed in the legendary plant have been spotted as far north as the rocky coastlines of British Columbia and as far west as the desert landscape in Nevada. Its influence on the ecosystems it arrives in spreads far and wide. But kudzu's real power is in the way it thrives in our collective consciousness, the way this vine with origins in East Asia has been swallowed up and remade into an American narrative, a southern storyline. It remains a symbol of a region it once enthralled, deployed by some as a way to "other" communities that they believe do not belong.

For everything from crops to shrubbery, in American popular culture, an invasive vine with Asian origins has long signified the end of times, known to smother all in its path with an embrace of certain death. To many, it is a curse of mortality driven by the harbinger of devastation to the natural world, an enduring legacy as the "vine that ate the South." For a select few, it has begun to morph into something else entirely. In its roots, a network of people sees a chance at redemption and an opportunity to rewrite a fragment of troubled history.

Set in the South, but starring characters spread out across a nation, this is a tale of cause and effect, as well as danger and downfall. It features a peculiar cast: an ecologist, a farmer, an architect, a chef, an influencer, a doctor, an author, a biologist, an entrepreneur, a forager, an entomologist, a historian, an analyst, a teacher, an artist, a scientist, and an engineer. The list stretches on, as long as an unbridled kudzu root trailing into tropical soils. All hail from different parts of the globe. All represent different cultures, languages, and perspectives. All seek different, often contrasting, outcomes. All are linked by one perpetually ascending, voracious vine.

From architecture teams using it as a building material in pursuit of a low-carbon supply chain, to clinical applications treating binge drinking and restaurants serving it as delicacies, these pages will bring to life the ways kudzu is being wielded across the US. By detangling the complicated dimensions of one country's relationship with kudzu, this story deciphers how the plant has evolved with the passage of time, as well as what we know about the impacts of climate change on its boundless presence. It bears witness to the remarkable ways public perception of kudzu has morphed—as the people living in territories overrun by the vine have bounced between embracing its abundance and fighting to destroy it.

When it comes to kudzu, the rolling waves of public opinion clash on one heavily personified plant. Rippling as they shift direction time and time again, cresting swells unveiling glimpses of mistakes and misconceptions, they are pointed transitions in-between the public and a sea of radiant green. With little understood about one of the most notorious invasive weeds in southern history, a lack of knowledge that compounds with the heightening polarizing rhetoric on otherness in America, there's never been a more opportune time to look more closely at an emblematic vine a long way from home. What lies between these pages is the unbelievable record stemming from kudzu—the lives, traditions, and communities it brought together and those it seamlessly wrenched apart. It is a chronicle following one weed's culturally chaotic journey from southern celebrity to American castaway and the people behind its undoing.

This is a narrative of belonging, racial ambiguity, outsiders and insiders, and the path from universal acceptance to undesirability. It is a deeply reported tale of mystery, an investigation into the past, present, and future of a quintessential plant. This is a story of sacrifice—an account of the ways a nation turned its back on those that did not belong. It is a saga of intrigue—a dive into the farthest reaches and darkest depths of the landscapes housing the many species we fight to control. At last, this is an ode to the earth around us, a quest for meaning in today's imperiled world. All found woven within kudzu's luminous, leafy vines.

A GRAND DEBUT

Somewhere off of Highway 64, along a series of winding roads in landlocked, rural southern Tennessee, linger the ghosts of a once-thriving dairy farm. For one little girl, the property served as a crucial ingredient to a blissful childhood: days blending into years spent devotedly exploring the fields and vegetation that were staples of those sprawling acres she once considered home. Now fifty-one, with a family and plot of land of her own, Beth Phillips sees how influential her origin story was. It was there, on that farm, playing among roving pastures and streams of cerulean waters that glittered like dewdrops in the rays of dawn, that she got her first introduction to the natural world and all its bountiful glory. For more than a decade, she has worked to embrace the promise of a wild, harvested American plant with its very own unusual genesis: kudzu. "I feel like I'm truly grounded in the earth," Phillips said, the southern twang in her voice softening, almost as if she were standing in a millennia-old place of worship. Her emotional connection with the cosmos comes through when she describes the soils, trees, and rocks blanketing the landscape around her. "Have you ever seen the veins in your arm or the way your lungs look when you breathe? The capillaries, all the little tubes—they look like the branches of a tree," she noted. "You can see the parallels, the sameness that you see in the human form." Her story with kudzu dates back several years, but kudzu's story in the US goes back much further than that.

It was in 1876—in the buzzing, dazzling environment that spoke of a nation in its glorious beginning—that the vine first landed on American soil, brimming with poise and promise. Kudzu was in Philadelphia for the first United States World's Fair, which cost more than $11 million to prepare for, all in honor of the one-hundred-year anniversary of the Declaration of Independence. A little over six years previously, a bill lobbying for a show-stopping event to honor the immensely important document was submitted to the United States Congress. In March of 1871 it passed, and plans were set in motion as the US Centennial Commission was formed. More than five years later, on May 10, 1876, President Ulysses S. Grant opened the Centennial International Exhibition, or the country's first official World's Fair. Upwards of nine million people—equal to almost a fifth of the US population at the time—came together in Philadelphia's Fairmount Park for the celebration. Also known as the 1876 International Exhibition of Arts, Manufactures, and Products of the Soil and Mine, this affair was a symbol of international power, prowess, and progress. Bushy-tailed and bright-eyed, the millions of visitors weren't just Americans making the trek cross-country. Journeying by train or by sea, people poured into Philadelphia from all corners of the globe to witness the splendor firsthand. Representatives from fifty-six countries or colonies—from Brazil to Japan—flocked to participate. Portraits on display at the National Gallery of Art depict black-and-white scenes where enormous buildings spread out over 285 acres of fairgrounds flaunted various architectural styles—everything from Moorish horseshoe arches to the symmetrical ratios of the English Renaissance. It was a juncture to be remembered—a moment to be etched, gilded and gleaming, into international history.

With more than two hundred buildings hosting the exhibits at Fairmount Park, only nine countries had pavilions, and one of those was Japan. Gently curved beams on an earthen-tiled roof fed into a carefully assembled two-story structure that served as the Japanese building. Designed by the architect Matsuo-Ehe, it relied heavily on fine-grain wood and shirked any use of iron in its construction. On display were the nation's prominent craftsmanship and detailed landscape design, subtly signaling Japan's shifting political priorities to an international audience. In control at the time was the Meiji regime, which decided that an elaborate production at the first official US World's Fair was a necessary step as it sought to remake the global perception of its nation. The display would serve as a grand way to reintroduce the world to Japan, which was in the process of industrializing to ensure its continued evolution. There, in the heart of Philadelphia, this painstaking rebranding would receive special, unfiltered scrutiny. The regime was prepared; accounts tell of how the

Japanese exhibit at the first World's Fair stretched across several acres, including an artfully constructed bazaar, pavilion, and garden—all sponsored by the
Japanese government and, according to historical records, built by Japanese
workmen. But the country's daring representatives hadn't made the expedition on their own. They had also brought along one nondescript, innocuous-
looking plant.

Bands of kudzu, which decorated parts of the showcase, were heralded as
a pretty porch ornament that could double as a soothing source of shade. Lush
greenery was in abundance in the first Japanese World's Fair exhibition stall,
one of which was the perennial guest.[1] Native to East Asia, kudzu was first introduced to Americans on that international stage. The immediate charm of
it lay in the singularity of its look: sheets of bristly, yellow-green leaves framing slender brown stems that spiraled up into the skies above. Kudzu likely
made an impression, but it wasn't isolated in its warm reception; the entire
Japanese bazaar stole the spotlight. An intricate craftsmanship style shone
through in the traditional designs that were built to illustrate the country's
modernization. Notable attendees raved about the Centennial exhibit, including prominent American architect Richard Morris Hunt. Hunt was reportedly
enchanted by the display; he later wrote it provided "capital and most improving study to the careless and slipshod joiners of the Western world."

On November 10, 1876, a brisk fall breeze swept through elaborate international demonstrations in Philadelphia held captive by intrigued crowds.
With it fled a final chance to witness the wonder of the first US World's Fair.
Six months after it had begun, the industrial exhibition shuttered its doors one
final time. At last, those who had embarked on an international expedition on
behalf of the Japanese government boarded the ship taking them home. The
vessel was loaded with artifacts and artistry that had stunned people from
every corner of the world. Among these tons of precious goods, ribbons of
kudzu lay, also awaiting something. Either a return to its native countryside
or an imminent demise. See, this wasn't kudzu's moment to splash onto the

---

1. Although plants from other countries were featured along with vegetation from the States
themselves in the grand glass building that was called the Horticultural Hall, the tempestuous
vine was not included in that secondary spot. See Fairmount Park Commission, "Catalogue of
Tender Plants Grown at Horticultural Hall, Fairmount Park, Philadelphia," Pennsylvania Horticultural Society, January 1906, https://doi.org/10.5962/bhl.title.115934.

trendsetting scene of the South—at least, not just yet. Although Philadelphia's Park Commission had no interest in letting the Japanese-style-frenzy sparked by the World's Fair go with the close of the exhibition—as they even imported a Buddhist temple gate to remain on the site of the Japanese bazaar—kudzu wasn't immortalized alongside the celebrated craftsmanship. Its preservation wasn't legally allowed. Along with all the other plants brought over to be featured in the fair, the weed that had traveled thousands of miles was destroyed. And with that, kudzu and its mysteriously beneficial properties vanished from an American vantage point. Gone, but not forgotten. It would return, in a tour de force that would rival all others. Falling under a scrutinizing spotlight that would only intensify, this reintroduction would help rewrite the narrative for thousands of plants to come. For kudzu would rise quickly into prominence, only to be chosen as the scapegoat, the poster child of invasive plants and the pinnacle of all that could go woefully wrong.

"It's been used for thousands of years in Asia," said Phillips. "Nobody does that here." More than half a century spent as a "lifelong resident of the South"— born in Tuscaloosa, Alabama, and raised in Tennessee—means she knows all about kudzu's tangled history in the US. The tale of one vine's tumultuous past and the perceptions of it remain ever-shifting in the present: stereotypes as deeply embedded as stubborn roots in thirsty grounds. Known by friends and peers as the "Kudzu Mama," which doubles as the name behind her brand, for the last six years, Phillips has lived in Rogersville, Alabama, where she has established a local reputation for her work with something many love to hate. She bills herself as a creator who "mingles with a vine with a horrible reputation and turns it into something useful and beautiful." Based out of her house, Kudzu Mama Designs is Phillips's business and digital storefront where she sells her unique handcrafted designs using kudzu. A long-running passion project turned professional odyssey, her love affair with the remarkable weed began when the artist was a stay-at-home mom. The otherworldly lure of the plant led to a serendipitous introduction.

In 2009, Phillips and her first husband settled in downtown Birmingham. Looking down onto their backyard in Southside was a sloping lot that connected to Red Mountain, its forested shelves grazing the clouds, parts of which were enshrouded in blankets of a single plant. The Appalachian beauty that soared to meet the skyline above was crowned in wrenching kudzu. "It

was spilling down into the neighborhood," she recalled. The property was situated on a dead-end street, positioned right by the leafy hill. "The kudzu was
growing right up against the house." By then, Phillips had a degree in visual
arts and experience working in an Atlanta art museum under her belt. Although she hit pause on those aspirations to homeschool her kids, she found
ways to incorporate plenty of exposure to a dormant creativity into their curriculum. Three and four at the time, her eldest daughter and son fixated on
the beckoning plant that spiraled down into their backyard playground like a
vertical bridge into another universe. That was all it took: Phillips began helping the children pull off leaves and gather roots to make wreaths out of kudzu.

It wouldn't be long before the little ones lost interest in crafting with the
plant, swapping their obsession with kudzu for another shiny new object of
their affection. But their mother found she couldn't move on as swiftly; that
first encounter molding kudzu into something tangible had lit a budding fire
within Phillips, a pulsing desire that unfolded like a rhythm to carry forward
creating with it. It was there, in her own backyard, that the artist found a
connection to the root of a plant that had dramatically melded to the exterior
of her home. From that moment on, Phillips found she couldn't stop herself
from making more with kudzu. It was as if they were destined, as if fate herself
had brought her to it. She kept on crafting—getting bolder and bigger with
her designs. Within a week, she had a seven-foot sculpture in the backyard.
She had incorporated wooden boards to serve as the base of an installation
that resembled a "crude" cross shape, while making a figurative structure in-
tertwining with the form fully out of kudzu vines. A long-time art enthusiast,
constructing with kudzu was Phillips's first foray into creating art of her own. "I
didn't know what I was doing," she admitted. "I just kept messing with it." That
continued until it morphed from a playful hobby to something meaningful,
something more.

Phillips's sculptures and figurines made entirely of the hulking being that
could be found haunting the countryside would eventually catch the attention
of a neighbor, who owned a collectibles store downtown and helped promote
her first show in 2009. As community interest in her work ramped up, she
started selling her individual pieces at other local craft shows. It was around
then that life got in the way of her work—so Kudzu Mama Designs went on
an undetermined, unexpected hiatus. "I went through a really bad divorce and
depression," she expressed. "I'd gotten away from it a little bit because life just

changed. I quit doing it." But the plant that first enchanted her into creating would eventually reel her back in. Over the next thirteen years, Phillips taught herself how to make functional objects with the "vine that ate the South." From start to finish, her creative process is labor-intensive; she pulls the vines she uses herself, then works with them while they are green, first drying them so they hold their form, before she seals them in a protective layer of polyurethane that produces a shellack, evoking a burnished finish. Creating a color variation, the sealant allows the vines to shimmer as if they are nature's jewels, enhancing the kudzu. "It's almost like the light plays off of it, too," she added.

Phillips is the founder and sole employee of Kudzu Mama Designs, where she creates, sells, and displays everything from light fixtures to plate chargers to kudzu fairies. All are built using a bristly, bendable plant. "Everything that I make is only made from kudzu," she said. Many of the light fixtures she sells are also designed using old, discarded lamps that she's found, taken apart, and repurposed. The idea to turn the weed into a source of light came from an artistic desire to embrace the shadows in her work and in her life. Although she focuses on light-related designs, the artist dabbles with kudzu-based sculptures, too. These range in their ingenuity, such as a sculpture of someone giving birth lovingly crafted from the vine's root, which Phillips noted was auctioned off for the women's studies department at the University of Alabama at Birmingham. (She cannot remember much about the person who bought it, aside from their desire to be an obstetrician.) These artistic celebrations of kudzu are on display in public galleries and private collections across the state. The list includes locations such as Home Grown Art Gallery in Sheffield, Grey House Gallery in Florence, and Kentuck Art Center in Northport. In the last few years, she has gotten creative about where she sources the items she adorns her kudzu creations with, finding materials everywhere from thrift shops to dumpsters. "I didn't have a lot of money to buy stuff," Phillips said. "So I always look to repurpose things."

The kudzu itself is sourced from a slew of spots nearby. There's plenty of it in Lauderdale County, Alabama, where the artist now lives. In the past, Phillips would stop and gather the plant she used from various towering patches, growing off of roadways, anywhere she spotted it. But Phillips no longer does that, preferring to avoid the risk involved—a burden that all women inherently bear when faced with navigating remote spaces. "I prefer to go to spots where I can just relax and I don't have to worry about somebody sneaking up on me," she noted. Of late, she gets it from three go-to locations. The first is the green spaces around Cypress Moon Studios, an event space in Sheffield—in neighboring Colbert County—right on the Tennessee River. "It sits below a cliff covered in kudzu," she observed, which blooms in abundance up against the studio. Her second go-to source for the vine is a friend's property over-run with it. The whole corner of his house is covered in the "scourge of the South." He's fought it every year, according to Phillips, who has now taken that job over, and for good cause: "I pull his kudzu for him." Her third treasure trove of kudzu is found tucked behind Florence, Alabama's Big River Studios, a recording space for local radio stations. Between those resource-rich spots, she has unfettered access to more of the weed than she could ever long for, a crucial ingredient to her artistic endeavors. At each of these chosen places, kudzu is plentiful enough that a penchant for avid collecting is not minimizing it but corralling it away from nearby properties. Despite all of her regular trips to collect the vine, it remains relentless in its abundance—and trying to creep back. "I'm keeping it from coming up on their houses. You can see that I'm making a dent that way," she said. Only recently has Phillips invested in Kudzu Mama Designs full-time. Her best year profit-wise was 2021, with the artist bringing in an estimated $10,000 in sales. A good portion of those funds has gone to travel—allowing her to bring her compositions to local and regional festivals and craft shows, her main wellspring of revenue. Phillips hopes to start experimenting again with bigger kudzu-based sculptures soon as she continues to immerse herself in her kudzu-based craft. "I'm trying to put out there to the universe: 'I need a house with a studio space and it needs to be somewhere in a kudzu field,'" she declared.

Raised Baptist, Phillips now identifies with Earth religions—with Pagan, Wiccan, or Druid being some of the most popular names practitioners call themselves—where elements of worship are reflected in her life's work and relationship with natural phenomena. This manifests in how she decides what to create with kudzu. If you ask the artist, she will tell you the choice is not her own; instead, it's a missive, sent to her from the very plant she is called to create with. "I'll just see a vine, or sticks do the same thing to me, tree branches, rocks and just things in nature, and I'll see faces in things," she explained. That's how her first figurine came to be; while gazing at the grooves and markings covering a knobbly kudzu root, the artist saw an image come together of a man. "I could see it. I mean, I can see it as clear as day right now. I knew what that had to be. And that's how it started." In the same way some horticulturists will tell you they "talk" to their plants, Phillips talks to kudzu. And it talks back. "I don't know how to really explain it. Except it's more of a feeling. It's not like I'm hearing words," she continued. Something instinctual guides Phillips while she works with the invasive plant. "It tells me which vines to pick. If it doesn't want me to pull it up, it's not going to pull up," she said. Her practice is freeform; Phillips doesn't follow a set strategy or approach in how she interacts with kudzu. She learned that the hard way. "If I do pull on it, I'm going to hurt myself every time. So I've just gotten where I go on the path of least resistance."

Now more than a decade into creating delicate, quirky designs based around one singular weed, Phillips hopes her work will raise awareness about the many ways the plant can be put to use. Everything she incorporates into her art signals a commitment to giving new vitality to the discarded; from broken lampshades salvaged from the trash to the hunted roots of kudzu. "There's so much that's wasted," she mused, "and that makes me sad. I don't want to be a part of that." There is a sincerity to her words that's almost tangible, as if any second now, a silhouette of a kudzu root will start taking shape in the space before us. Moments of silence follow her admission, conjuring an image of a human in careful, fervent companionship with the land. In a world where many seek opulence and exaltation, giving little thought to who or what they hurt along the way, Phillips's instinct to preserve the environment and provide for others—her two main life ambitions—sets her apart. She hopes that those who see her work will be inspired to look differently at the biomes around them, to treat the earth and its resources far better than we do. "I would like for my art to reflect that very much, because I strive for my life to reflect that," she noted.

Wielding words that offer a sense of optimism to those open to taking them in, Phillips aspires to leave as modest of an imprint as possible while honoring the kudzu—thrumming with power, purpose, and potential—all around her. "We're all connected. We're all stardust. We're all energy," she said. The artist whose work revolves around kudzu is one nexus in a community scattered across the South; strangers entangled by the pull of an enthralling vine.

"At our core, we are all the same," she said. "And I love that."

2

PORTAL INTO THE PAST

A little quirky and a whole lot passionate, Lauren Bacchus can talk about kudzu and how it interacts with the world around it for hours. She is not dry or monotone about what she loves about it either. Unfiltered and lively, Bacchus comes off as effortlessly charming as she jumps from one train of thought to the next. The only time she stops her tirade of information about the plant in question is when asked what lessons she has learned from working with kudzu. Responding in a tone of reverence, this line of thought trips Bacchus up. "Gosh, it's—I'm such—it can go so many different ways," she said, falling silent and still for a while, lost in an array of possible responses. "I'm sorry, it's a lot," she explained. "It's so important and so overwhelming that I'm frozen." Her answer matters this much to her because for Bacchus, kudzu matters. It is the green beating heart of her personal kingdom.

Born in Twentynine Palms, California, Bacchus and her family moved to Pennsylvania, then Texas, before settling down in Davidson, North Carolina, when she was nine. It was there, playing in a field by their house bursting with patches of boundless kudzu, that she first encountered the weed. "My sister and I would make up different animals from the kudzu vines from the trees," Bacchus said, recalling how they would use the kudzu to further their childlike performative games. She compares it to scrying, an act of divination that involves gazing into a surface in hopes of seeing what's to come. "They would turn from dinosaurs to elephants. Kind of like clouds, but [with] kudzu." The family also lived near another parcel of land devoured by kudzu which Bacchus would use as a destination point—routinely throwing sticks in the thicket for their dog to fetch. "As a kid, it's mysterious," she mused, remembering how the perennial called to her from the start.

A nine-year-old Bacchus had it right; there is something inherently mystical about the yellow-green stems and sometimes heart-shaped leaves that signify a force capable of overwhelming all in its path. Back in the early twentieth century, that initial mystique became something increasingly complicated. Over a period of mere decades, the plant that once inspired annual festivals and fan clubs morphed into a bothersome enemy within southern culture. Ever since it first ensnared Americans, kudzu has played an influential role in US history—setting the stage for an entire country's somewhat fickle relationship with invasive plants as a whole.

Formally introduced to the US through a Japanese exhibit at the 1876 World's Fair in Philadelphia, kudzu was first heralded as a decorative vine, with leaves so prominent and large they could be used to shade a home. For seven years, the ornamental living object was little more than a mere memory for most Americans. Then, in 1884 and 1885, kudzu sprang up again at the World's Industrial and Cotton Centennial Exposition in New Orleans—a lavish event featuring gigantic wooden structures and broad, lighted paths that once again brought together plants, art, and industrial achievements from around the world.[1] More than a million visitors showed up to see what the National Cotton Planters' Association had sponsored. The weed, not yet a staple of the South, re-emerged at this very juncture, as an international emblem of sorts. For those in attendance, records suggest the presence of a plant native to East Asia demonstrated that the US was interested in further developing its formerly estranged relationships with non-Western nations.

Behind the scenes, however, enterprising residents had found a discreet way to keep one compelling plant circulating in the States. From 1862 to 1874, US consul and horticulturalist Thomas Hogg Jr., an American working in Japan, routinely shipped the vine to his brother in New York City. Hogg's family made money off of the vine by selling it in bulk at their garden center. From there, nearly all recorded mentions of kudzu in the US vanish as if it were a ghostly being summoned only at gatherings of international proportions or popping up in a handful of isolated cases, haunting a sparse network spread

1. The site of the first international cotton exposition was rife with tragedy. Three hundred acres of what had once been a plantation housed the event—land forever entrenched with buried examples of the darkest depths of human atrocity. See "The 1884 Cotton Expo and New Orleans' First Case of World's Fair Fever," *Nola,* May 17, 2017, https://www.nola.com/300/article_21fc06f9 -a1f5-56f8-8440-dee682805dfe.html.

out over a nation. It would not be until decades later that kudzu would make its legendary reappearance, re-emerging into the American spotlight with a certain panache we would later learn was the steady thrall of one charismatically abundant plant.

Its seeds were the first element of kudzu to receive mass appeal in the West. The late nineteenth century saw an uptick in imported kudzu seeds to the United States, sold and marketed for the purpose of growing a purely ornamental vine. By the close of the century, the seeds became available through mail-order catalogs. One thing led to another, and kudzu began to be used sporadically across the region as a means to hide houses from the heat with style and ethereal flair. In the South, it offered a dazzling, rapidly growing solution to a persistent problem: sweltering heat. During that period, the weed would be most commonly used to shade porches or courtyards. Providing an escape from the scorching heat was one way kudzu made its initial mark; another plus was the grape-like scent the vine's blossoms produced.

Around 1902, one influential Chipley, Florida, resident and businessman, Charles Pleas, began touting kudzu as an inexpensive forage for livestock. After hanging kudzu on their front porch as a form of protection against the state's unrelenting temperatures, Pleas and his wife Lillie had discovered that the plant appealed to their farm animals. This led them to found Glen Arden Nursery and eventually start a mail-order catalog capitalizing on it. Pleas would later publish a pamphlet in which he wrote enthusiastically about the vine's utility. In 1907, kudzu was included in an agricultural exhibition in Jamestown, Virginia. A decade later, the Alabama Agricultural Experiment Station launched research into the plant as a source of animal feed. By 1920, conservationist John Rigdon got involved, kicking off a campaign for kudzu backed by the Central Georgia Railroad, his employer and one of the most vital railroads in the South at the time. Rigdon was drawn by the possible economic benefit of shipping the weed, hoping that kudzu hay would open a new venue for the company's freight demand. This sparked an advocacy project that gave away free kudzu seeds and pamphlets to farmers across the state, promoting the vine's use for livestock. It would set off a domino effect, one that ascended all the way to the federal government, where credit for the origin of the US Department of Agriculture's interest in kudzu belongs to one scientist. American botanist David Fairchild doggedly experimented with planting kudzu roots imported from Japan, followed by seeds purchased from a local nursery, on his property in Washington, DC. According to his written accounts, the former method met with zero success while the latter took off. By then, his pursuit of kudzu had caught the attention of colleagues at the federal agency, which also began to test the application of the plant—buoyed by agricultural anecdotes that had begun circling around its utility as a cheap source of forage for hungry, grazing animals.

In 1933, the federal government established what was then called the Soil Erosion Service. It would be renamed the Soil Conservation Service in 1935 and then the Natural Resources Conservation Service in 1994 (which it remains known as today). Congress tasked the agency with reducing soil erosion caused by poor farming practices, particularly in the South, where dry land was most common. Farming was the occupation listed for almost one-third of people living in the US, according to 1920 census data. After all, this was still a predominantly agricultural era, where farmers and industry workers across the South shared a common foe: human-induced degraded soil. Historical records date back to 1909, when Hugh Hammond Bennett, an inspector at the US Bureau of Soils, mapped and reported the first soil survey that identified soil erosion in the South's plantation regions. There were a number of causes for this dry soil. A leading factor was the South's legacy of slavery driving its agricultural systems. Brian Barth at *Modern Farmer* wrote in 2016 that the "Black Belt" first received that name as a nod to the region's rich topsoil but that the term eventually came to refer to the slave labor that fueled production of two-thirds of the world's cotton. A surging demand for cotton and forced, free farm labor led plantation owners across the Black Belt to overuse and neglect the soil, employing farming practices that severely depleted it. According to Barth, the compounding pressure of the introduction of the boll weevil across the South "decimated" the cotton plants, and in the 1920s, the "industry was wiped out nearly overnight."

Enter kudzu. Where sheets of it grew, even the most depleted, dehydrated soil was said to transform. The vine's leaves and root system were later found to have the ability to interact with the low pH of dry soils, with the potential to "enhance soil fertility" and "control soil erosion," per a 2016 article in the *American Journal of Plant Sciences*. Another asset of the perennial was its remarkable rate of growth; the word quickly spread that it could grow up to a foot a day at the peak of its growing season, with vines that could shoot up dozens of feet. All elevated by the plant's alleged ability to successfully grow in eroded land. That was all the federal government needed to know—once they found a solution to a problem that had led to significant agricultural and economic losses, they saw nothing but dollar signs, sprouting off of canopies of trailing kudzu. From there, kudzu's climb to acclaim was boosted by several government-sponsored moves to thoroughly market it. The first major attempt was in the 1920s, when the Central Georgia Railroad invested in thousands of kudzu seeds and gave them to farmers to plant. Met with skepticism by the people they were pushing kudzu on, the campaign did not have much success, but US Department of Agriculture researchers didn't give up; the agency still harbored high hopes about kudzu and its rapid growth and soil nutrient stabilizing capabilities. The year 1935 marked the first widespread efforts to mitigate natural soil erosion by planting kudzu. Touted as a magical cure for diminished soil, the "mile a minute" vine rose in prominence on the American agricultural scene. And just like that, the invasive would become a household name.

In the 1930s and 1940s the kudzu fervor reached its peak as reports spread quickly about how the plant gave new life to unsalvageable ground. This messaging was pushed by the Soil Conservation Service in its crusade for improved agricultural ecosystems—and of course, economic gains—in the South. The vine was also marketed as cheap, effective forage for livestock. By this point people were racing to plant kudzu, especially when they heard they would get paid for planting it.

Accounts differ on how much kudzu was given to southern landowners by the Soil Conservation Service, but an article last updated in 2022 by invasive plant specialists at Alabama A&M University and Auburn University puts the number at 85 million seedlings. Farmers were told kudzu would do two things: reduce agricultural erosion and boost nitrogen levels in soil that desperately needed revitalization. The US government offered incentives to farmers and agricultural workers; it paid $8 per acre, roughly the equivalent of $169.60 today, to anyone who planted their land with kudzu. Thanks to that mass marketing movement spearheaded by US officials, the economic potential of kudzu became widely known. And what began as a means to achieve optimum relief from the heat evolved into a heralded remedy for the most pressing agricultural and economic challenge facing the South. But the Soil Conservation Service was not alone in spreading pro-kudzu propaganda. The agency was joined by the Civilian Conservation Corps, a voluntary work relief program established by President Franklin Roosevelt that ran from 1933 to 1942. It deployed unemployed, single men to help mass-plant kudzu across the Southeast in a program intended to help hoist the United States out of the unrelenting throes of the Great Depression. Farmer, journalist, and radio show host Channing Cope was another source intent on spreading the word about one handy vine. It was then that the weed was morphing from an agrarian icon to a cultural one. Cope jumped on the kudzu bandwagon, dedicating most of his public platform to writing and speaking about the miraculous properties of the plant, which he crowned as a divine answer to soil erosion. He founded the Kudzu Club of America, which would go on to include roughly twenty thousand members, earning him the nickname "Father of Kudzu."

By the mid-1940s, about three million acres of kudzu had been planted in the South. In the lead-up to this reckoning, the rise of industrial America had already unfolded. By the early 1900s, farmers were abandoning their lands in droves to relocate to shinier, newer cities—where the prospect of riches and excitement beckoned all who dared. City populations nationwide ballooned between 1800 and 1900, with records revealing a total migration boom of roughly 15 million people, a mix of immigrants and residents of rural America lured by industrial growth and job opportunities. Nearly 40% of US townships faced dwindling populations between 1880 and 1890 as people ditched farms and increasingly competitive livelihoods battling a technological revolution for booming, budding urban terrains that promised steadier paychecks. The trend continued as Americans packed up and left unmonitored lands where kudzu seedlings had been planted, conditions ripe for overgrowth, behind. With no one around to manage it, the vine began creeping unchecked across the countryside, covering every crop, plant, and object in its path.

It would take crops being overtaken by kudzu for southern farmers and others working in agriculture to begin to panic. It was in the early 1950s that the American public's understanding of kudzu once again shifted. At that point, anecdotal complaints about the weed had already started to surface, with emerging reports suggesting that people were struggling to get rid of the plant they had recently harvested for better soil and a quick buck. The word was out: The vine's rapid spread across the South wasn't solving problems in agricultural production, but causing them. Farmers were discovering that where the plant went, destruction followed. Wrapping its sheets of leaves so tightly around vegetation that it shut out their access to the sun, kudzu was disrupting ecosystems, killing crops, and overtaking farms, fields, and other open habitats. It also was spotted enveloping telephone poles and power lines. In some pockets of the South, kudzu had even crept over railroad tracks, leading to train delays and accidents. The purportedly "uncontrollable" rate at which it spread was powered by rising temperatures in some parts of the South, which spurred more rainfall and longer growing seasons, per 2022 National Oceanic and Atmospheric Administration (NOAA) data measuring changes in annual average temperatures across the contiguous US states since 1901. Given that the plant originated elsewhere and was widely placed in a brand-new landscape, there were no native insects on hand to eat it, nor was it flourishing in a native ecosystem that could curb its rapid rate of growth. A new narrative began to form about a weed that had been worshiped as a miracle, one that centered it as a source of chaos and a target of condemnation. All of these elements combined to boost the geographic spread, and dovetailing cultural downfall, of a once-beloved vine.

The federal government responded to complaints about kudzu gradually, displaying the same urgency as a pocket-sized animal caught in a slow-motion sensory world. It initially removed the noxious species from the list of cover plants permissible under the Agricultural Conservation Program in 1953, and nearly a decade later, limited its recommendation of kudzu to only nondeveloped areas. It would take another eight years before the US Department of Agriculture classified it as a common weed and more than two decades before legislators in Congress would vote to put kudzu on the Federal Noxious Weed list in 1998, under the Federal Noxious Weed Act. It would go on to be removed from that list in 2000. As of 2022, even though it's not federally recognized as a noxious weed, kudzu remains classified as one in thirteen states.

To some degree, the vine that reportedly "engulfed" an American region did manage to find its way back into the public's good graces in 1985, when some alternative uses for it began to surface. But that minor villainous redemption arc didn't last. By that point, kudzu's reputation as a perilous pest had been long ingrained into sociocultural consciousness. The decision had already been made; kudzu had been cast as an antagonist, and it played the part all too well.

This peek at its history makes one thing obvious. Ever since it was first ushered into the US, kudzu has had a long, nuanced relationship with the Deep South. Beyond American borders, the renowned vine has thrived in sprawling countrysides across the world for millennia. It's a plant that can be traced through thousands of years, mentioned in ancient Chinese literature dating as far back as 500 BC and found within Japanese records from the eighth century. A 1973 journal article by Chinese botanist Hsuan Keng in *Economic Botany* identifies the earliest-recorded mention of kudzu in *Shih Ching*, or *The Odes of Poetry*. The oldest existing collection of Chinese poetry, *The Odes of Poetry*—otherwise known as *The Book of Songs*—is believed to be the first anthology of Chinese poetry. It was compiled by Confucius, the philosopher whose teachings deeply influenced East Asian governments and views on morality. Made up of 305 works dating from 1,000 to 600 BC, it includes a single mention of kudzu in a poem titled "Collecting Kudzu Vine," translated to English below by modern-day poets Tony Barnstone and Chou Ping:

Collecting Kudzu Vine

He went away to collect kudzu vine.
One day's absence
is as long as three months.
He went away to collect wormwood.
One day's absence
is as long as three seasons.
He went away to collect moxa.
One day's absence
is as long as three years.

In *The Book of Kudzu: A Culinary and Healing Guide,* originally published in 1977, authors William Shurtleff and Akiko Aoyagi traced the likely first recorded origin of kudzu in Japan to Manyoshu, a collection of poems compiled in 600 CE. Translated as "Collection of Ten Thousand Leaves," historians have since credited the anthology to the poet Otomo no Yakamochi, who is said to have put together the oldest-existing collection of Japanese poetry in 759 CE.

On Blossom

The kudzu vines upon the moors
Flutter in the autumn wind
As it blows
Upon Ada plain
The bush-clover blossoms scatter.

Fast forward a millennium, and there are a select few who still worship the properties of a weed that serves as a portal into the past. Bacchus, who leads Kudzu Culture, is one of them. In 2017, she joined forces with fellow kudzu aficionados in the area, Zev Friedman and Justin Holt, who had launched the group the year before.

It's 2011, and Holt and Friedman are running point on a project to grow a garden on an abandoned lot overrun with unruly vegetation, where they need to remove the roots of a vine called kudzu from the ground. While digging, they discover that the soil where the roots are growing is rich, fluffy, and dark. The source of that, they would soon realize, was the kudzu. Later that year, the two would be astonished to witness the formerly rundown lot transform into an abundant, bountiful garden. And that was all they needed to see. The duo of permaculturists went all in on the vine—first launching Kudzu Root Camp, an annual event hosted at Friedman's parents' kudzu-overrun land in western North Carolina, before recruiting Bacchus to form their trifecta, and eventually serving as an official resource for locals curious about the plant's utility. Over the last decade, Kudzu Root Camp has grown into a regular affair, usually attracting two dozen attendees to biannual workshops, where they learn how to harvest and incorporate the weed into food and medicinal applications. From selling harvested kudzu to local kombucha companies to shipping raw roots through Etsy, one of the group's ultimate goals has been to build a kudzu-based product line in the Southeast.

One could argue that Kudzu Culture serves as a modern-day kudzu fan club: a community of people spellbound by the power within a plant. Within those ranks, a web of people has become connected by their shared crusade—working to resurrect the nation's idolization of the visually striking icon many love to hate. In recent years, they have been zeroing in on new projects like on-farm experimentation. "Harness by Harvest" teaches farmers how to prevent kudzu from having invasive growth patterns by intentional harvest and interaction (i.e., harvesting blossoms to hinder seed pod development, or to mitigate spread to tree lines and crops, and maintaining kudzu stands to be in ecological harmony). This is still a work in progress. Other projects include "Community Through Kudzu," an educational greenway maintenance initiative; "Vine to Cloth," a collaborative fiber-research project; and an ongoing "Invasive Plant Fiber Study Group," in partnership with Fiberhouse Collective and Local Cloth in Asheville, North Carolina, an education center and nonprofit supporting sustainable artwork.

One day, Kudzu Culture aspires to serve as an educational resource for all things kudzu in the US. "The goal eventually is to be a cooperatively owned network of kudzu," said Bacchus. They hope to do that in a number of ways—by harvesting and selling kudzu, hosting community events, and being a digital resource for any kudzu needs. Online products include seasonal fresh kudzu root, sold during the winter because "that's the time that has the most active constituents"; culinary-grade tea, which Bacchus says is typically made from kudzu starch, that some use as the basis for an elimination diet ("If they're having issues then it's like, 'Okay, well at least sip on this kudzu'"); and shredded dehydrated root, which is marketed to herbalists but also as a culinary application ("To make drinks and other things," she explained). All of the kudzu used in the products they sell is gathered from "high-elevation mountain sides" in western North Carolina, which Bacchus says are intentionally handpicked as places least likely to be exposed to runoff and ground pollution.

Just outside of downtown Asheville is the neighborhood of Montford. Once it was home to approximately half of the city's Black residents—until diverse neighborhoods were torn down and built over during an expansive urban renewal program. It's a very expensive community now, according to Bacchus, who cites the area's troubled history of gentrification. And there's kudzu all over the greenway there, an overabundance of an unwelcome plant that has spurred a sense of action by both the city and the state. Early last year, Kudzu Culture partnered with the University of North Carolina, the city of Asheville, Urban Forest Alliance, the Montford Neighborhood Association, and the North Carolina Department of Transportation (NC DOT) for an event where they taught a classic management tactic of kudzu: hand removal of kudzu crowns. Bacchus says the typical method for kudzu management is spraying herbicides, but without incorporating hand removal, this is largely ineffective. "See the DOT goes in, sprays herbicides where they see kudzu, and then they don't go in and hand-remove anything," she said. But without hand removal, the process fails. According to Bacchus, the dead matter creates a lattice for the next year's growth to grow on, which she considers "just a waste of money and resources." For their first event, the Kudzu Culture–hosted Montford project attracted some surprise guests—mostly "super fit" older ladies. "I was like, 'Wow, y'all are really liking doing the 'mattock' stuff," laughed Bacchus. "They were in their seventies!" For Kudzu Culture, Bacchus is also working on what she calls a "mobile processing lab"—which will eventually help achieve the organization's goal of establishing a kudzu-based product line. Step one for the team is to educate local farmers who have kudzu growing on their lands. The end goal is for those farmers to go through Kudzu Culture's testing process before becoming an approved kudzu harvester using a certified harvest site. "And then we will aggregate," said Bacchus. This means they'll source the kudzu they use for their products from local workers at the approved spots. In that scenario, they'll be able to pay farmers, approve harvesters, and green-light harvest sites.

Someday, not too far in the future, she sees this expanding to become a work opportunity for felons in the Asheville community who otherwise struggle to find employment, which a 2021 Bureau of Justice Statistics report found is a nationwide issue. "This is really powerful work for people to learn about, because the metaphors of the plant and being seen as, by some, as having no value, but then you learn that they have value," she noted. Developing a program that helps provide employment to a segment of the population that disproportionately needs it is crucial to Bacchus, who wants to form a cooperative-owned model to engage people who can't find work in other ways.

Kudzu Culture became a 501c3 in 2022. Bacchus is the interim director and the only paid employee; she runs point on essentially everything. She is ridiculously busy—flitting from all the admin that comes with operating an organization to coordinating community events and initiatives. When we last spoke, she was even trying to recruit for Kudzu Culture's first-ever board—all while working other jobs, like cleaning Airbnbs, to pay the bills. "I basically am an artist who has been living in Asheville for fifteen years. I've been getting priced out of town, and I have to string together a million jobs," she said. One of those jobs involves cleaning rental units in the area where she lives, just outside of the city. "And I'm obsessed with this kudzu project." Bacchus is determined to turn her part-time position running Kudzu Culture into full-time employment doing what consumes her. She's trying to find as many grants as she can to sustain herself, a resolute work in progress. "There's no stopping. I can't stop the kudzu work," she chuckled, part exasperated and part exhilarated by her own devotion to the vine. "I need to be doing this work. So I have to figure out how to make it my job."

There's a philosophical Japanese concept that Bacchus stops our conversation for a moment to explain. It's called *Ikigai*—something that gives us a sense of purpose, a reason for the lives we lead. "It's like the French *raison d'etre*, a reason for being," she said. Bacchus sees working around kudzu as her personal form of *Ikigai*. It's not just about fulfillment though, she tells me; it's about your vocation—getting paid for what you're doing but also doing something important on behalf of others. "[It's] important for the environment—with the world, yourself. [It's a] feeling of interest and love." Her adoration for a plant native to countries thousands of miles away can be traced back to her own curiosity for cultures different from her own. With a background in fiber arts and paper making, Bacchus dropped out of college while traveling around East Asia. Her interest in Asian papermaking began during the year she lived in China. And she remembers drinking kudzu without even realizing it while touring Japan's scenic countryside. It was in the form of *kuzu-yu*, a traditional Japanese gooey drink. In Japan, repurposing the vine in culinary and medicinal applications is as commonplace as consuming oranges to boost vitamin C in the US. Bacchus has since rooted herself within an international community of kudzu fans—people across the world who also cherish the weed, just like she does. Relying on Google Translate, she keeps the Chinese and Japanese characters for the plant stored on her phone and searches with them on Instagram to find fellow vine lovers beyond American borders. "I have conversations and messages with really cool, younger people who are like, 'Oh my god. You love kudzu. That's awesome. I love kudzu,'" she continued.

Bacchus has big dreams—visions of a future that revolves around one tenacious plant. Right now, she's living in a camper at what she's deemed "the Airbnb farm." It's a site just outside of Asheville where tourists flocking to the city come to stay—a cycle of visitors from across the country, strangers whom Bacchus cleans up after once they leave. It's not long-term; she tells me she is saving money for a camper of her own—a haven on wheels that will allow her the full autonomy she craves. A call for a different life beckons her—one where she can live surrounded by a plant she just can't stop thinking about. "Stop cleaning Airbnbs, go live on the kudzu mountain, build myself a little tiny home made of kudzu composites," Bacchus said, describing her dream. One day, the kudzu aficionado hopes to live on a 20-acre plot in Marshall, North Carolina, that is carpeted in sheets of kudzu. It's not too far off from reality. Dense coats of the perennial drape over a cluster of mountainsides just a couple dozen miles north of Asheville. Her heart is set on that spot. Having spent most of her formative years in Appalachia, Bacchus already considers herself part of "the mountain people": those who are at peace with living in higher-altitude fields and lands overrun with wild plants like kudzu. "There's something about being in the old mountains. That's where I feel the most at home," she confessed.

She paused, her voice softening, before adding wistfully, "I'm saving money for a camper to move to 'The Kudzu Land.'"

3

KUDZU VERSUS THE SOUTH

A 155,000-acre expanse of forest thriving with diverse habitats can be found in northern Mississippi. Deep within the woodland is a sea of impenetrable kudzu. Fields and clusters of trees are completely swaddled in it, conjuring images of alternate worlds where pixies dance playfully in the shadows cast by the egregiously vast plots of land and pay tribute to sweeping leafy temples. A natural microcosm that comes completely and breathtakingly alive by kudzu. It's there at the Holly Springs National Forest, in the midst of clusters of mighty hardwoods and regal pines, that Gina Profetto first established a groundbreaking research project. After three years of navigating the forested fray, the overwhelming visceral response evoked by unbridled sheets of the vine has not lost its luster for the scientist. "Looking at a nice kudzu patch of acres of just green flush," said Profetto, "it's a sight to see."

Like a migrating leatherback turtle, known for swimming more than ten thousand miles in a year and setting the bar for one of the deepest recorded dives made by marine animals, kudzu is dramatic within the world of invasive plants. Overtaking other flowering and nonflowering vegetation, the vine devours indiscriminately, with little warning but ensuing fallout. It wraps its layers of bristly leaves so tightly around its prey that it kills native plants in the ecosystems it enters, crushing them with its weight and shutting out their key source of sustenance—the sun. A large, often trifoliate-leaved, semi-woody perennial vine that can blossom, kudzu can be spotted roaming southern roadways, forest edges, and farmlands. Left unchecked, it blankets everything from trees to buildings, engulfing immobile fixtures like a sea of fireflies swarming cattails in a pond. A member of the *Fabaceae* (legume) family, kudzu has approximately seventeen recognized species belonging to the genus *Pueraria,* whose name honors the Swiss botanist Marc Nicolas Puerari. All are native to countries in East and South Asia, including China, Japan, Korea, Thailand, Vietnam, Indonesia, Malaysia, Papua New Guinea, India, Taiwan, and the Philippines.[1] Kudzu is known as *kuzu* in Japanese and *ko, ko-shu,* or *Gegen* in Chinese.

At first glance, three broad leaflets make the vine look like a shoo-in for poison ivy or greenbrier. Upon closer inspection, spiraling stems and twining brown shoots give it away as it coils and twines through a perpetually ascending climb. Another feature that distinguishes it from its similar-looking counterparts is discreet, yet sporadically abundant, blossoms. These less-than-an-inch-long flowers typically pop up in late July through early September. The little blossoms are highly fragrant and more frequently found in ascending kudzu vines—sun-drenched, draped over trees, and climbing skyward. Blossoms are rarely spotted on kudzu with minimal exposure to daylight and elevation. They are noticeably absent from the vines that crawl around at ground level. The buds imitate the rainbow; kudzu's pea-like flowers can spring up as red, purple, magenta, pink, or white.

One thing that makes kudzu so conspicuous is its alleged rate of growth. The tap root is said to grow a diameter of more than 18 cm and extend in length about 6½ feet. Starchy, tuberous roots are a staple of the weed. Although reports vary, several suggest that the plant's thick, crawling foundation can weigh as much as three hundred pounds, and a single root can spawn up to thirty vines, each one capable of reaching a depth of roughly one hundred feet. In a single day, its vines are reputed to grow nearly a foot. Mysteriously, these reported figures vary and appear largely anecdotal. What's definitive is that kudzu flourishes in disturbed and degraded areas—roadside ditches, gullies, farms, rights-of-way, and the edges of forests. Nonforested lands are where the perennial thrives. Similar to marigolds or black-eyed Susan flowers, kudzu does best in full sunlight, although it does also persist in partial shade. But

---

1. A handful of reports also suggest that kudzu is native in parts of Australia; however, these accounts are conflicting, with some suggesting that the vine has instead been naturalized there. See Ka Sing Wong et al., "Kudzu Root: Traditional Uses and Potential Medicinal Benefits in Diabetes and Cardiovascular Diseases," *Journal of Ethnopharmacology* 134, no. 3 (2011): 584–607. https://doi.org/10.1016/j.jep.2011.02.001.

much about it remains an enigma. An assistant professor of biology at Abraham Baldwin Agricultural College, Profetto is quick to tell you that. After devoting five years of her postgraduate career at the University of New Orleans to studying kudzu, she knows all about the nebulous weed. In 2021, she published a paper in the *Journal of the Torrey Botanical Society* on the vine's ecological impacts in Mississippi.

Born and raised in Syracuse, New York, Profetto didn't have the slightest inkling of what kudzu was until she moved to Louisiana in 2018. In true millennial fashion, the twenty-seven-year-old credits Google with her introduction to the imposing vine. "I had no idea what the flora looked like in Louisiana," she recalled. But her move meant she suddenly found herself dealing with a completely new category of vegetation. "It really just took a few Google searches to realize that kudzu was the main invasive species in the South," she added. The scientist came to the topic fresh and chose to focus on it for her dissertation, which would lead her to spend years wading through fields overrun with sheets of the mesmerizing entity. Along the way, Profetto devoured the plant's backstory, trying to get a sense of the research that had been already done about it. Something struck her early on. A 2004 article published digitally in 2010 in *Critical Reviews in Plant Sciences,* which incidentally was one of the first studies she came across on kudzu, included a line that needled its way into the young researcher's mind. Study authors Irwin Forseth and Anne Innis wrote that more quantitative data is needed to understand the effect of kudzu on native species and its ecological impacts—as existing reports were drawn from narratives, not from fact. "Official campaigns to eradicate and control the vine continue unabated, even in the face of criticism that the techniques used for eradication are 'costly and labor intensive' and that much of the research on kudzu's alleged ecological devastation is anecdotal and lacks any quantitative backing," the paper noted. Just forty-five words near the end of the abstract imprinted into her brain a beacon that pointed to something missing in our collective understanding of kudzu. "We know that invasive plant species beat out native plants, but nobody was actually quantifying it. Nobody was looking at that," Profetto said. "It was just *assumed* it was," she observed, remembering how baffled she was that no one, to date, had successfully tried to back these anecdotes up. It was 2016 when she came across that stunning statement, and after digging into it further, was even more bewildered to find that a decade later, the sentiment still rang true. The then-twenty-one-year-old dove into the unresolved task with fervor, eager to be the one to either prove this widely believed assumption about kudzu right—or demonstrate the ways in which it was woefully wrong.

It would ask a lot of Profetto, the work being more than she bargained for. Voyaging into the world of scrutinizing the ecological consequences of a vine would send the fearless scientist into the Holly Springs National Forest, where she would brave rain, sunshine, ice, and snow in nine expeditions. Season after season passed while Profetto steadfastly moved forward with her work. The grueling nature of it was, at times, too heavy a burden to bear. "If you have spent a July summer in Mississippi, hiking up a hill in a ten-foot-tall kudzu patch, with God knows what's crawling underneath you, it takes something out of you," she said. All in the name of research, in a quest for clarity for one severely misunderstood vine. While there, Profetto identified sites that had previously been treated with different kudzu management techniques, from prescribed burning to herbicides, and compared the species presence with sites where no kudzu, nor management strategies, had been spotted or applied, as well as a site where kudzu was running rampant and no methods had been attempted to rein it in. Making three visits to these divergent sites in the summer and fall every year over the course of three years, Profetto regularly collected samples of every herbaceous species she saw in the plots, took them back to her lab and processed them, and at last identified what she had collected. Using pictures of each quadrant of her plots, she relied on an image-processing system to calculate the percent cover within the quadrants of the specific herbaceous species. For the woody species measured, she identified the species in the field and measured the density of the coverage before sorting out the different classifications back in her lab. In order to compare the measurements of species' presence, richness, and density in these plots, Profetto took a closer look at vegetation types impacted by kudzu, comparing grasses with herbs with vines. The questions centric to her work were: "What's going on with the vegetation? What are the differences between them? Is kudzu actually out-putting native vegetation?"

Over a total testing period of 1,095 days, Profetto collected thirteen woody species and thirty-nine herbaceous species from the understory, the layer in a forest below the bottom of the canopy made up of a mixture of smaller trees, shrubs, vines, herbs, grasses, and saplings. Her questions led to unexpected answers—and more questions. "We can say for certain that yes, kudzu did significantly alter the native plant community characteristics in the invaded sites compared to the control sites. [The] species richness of all plots, the percent cover of understory species and the density of woody species . . . were all significantly reduced," she said. But then they got into some specifics of looking at the shade tolerance of species. What did kudzu affect more—grasses, herbs, or vines? Profetto and her team found that it doesn't matter if the native plants kudzu overtakes are classified as shade tolerant or intolerant; those characteristics don't affect its persistence on understory or woody species. "It is not biased," said Profetto. "It'll outcompete you every single time regardless of tolerance." The scientist found this especially surprising because the vine's main way of outcompeting native species is by taking away their access to sunlight. It doesn't differentiate between species that are shade tolerant or not, a finding that Profetto said "blew us away."

Beyond that, they found no evidence that kudzu favors weedy versus nonweedy species. "Kudzu does not care," she concluded. Oddly enough, when it comes to other vines, kudzu did display an unusual pattern of resistance—it didn't reduce the persistence of vines as much as grasses and herbs. Of the six native vine species found across the testing sites, five are persisting, untouched by the prolific weed. "It went after the grasses and shrubs. It didn't really care about the vine species," Profetto declared. What their research revealed also debunks "Invasional Meltdown," a theory that dates back to 1999 alleging that the introduction of invasive species to a native landscape leads to the future establishment of other subsequent invaders. According to Profetto, there was no evidence that the presence of kudzu translated to other opportunistic weeds and species taking over the land. "It was all of the same species that we saw in the control sites," she said. "It was just that they were decreased."

One of the driving factors behind kudzu's overabundance is the nature of the spaces it invades. Kudzu can survive almost anywhere, in full or partial sun, but it thrives in disturbed terrains where daylight is abundant: at the edges of forests, on abandoned farmland, along roadsides, and throughout fields. Casting a substantial layer over any landscape it enters, the herculean plant inadvertently smothers everything from small plants to trees, throwing a blanket over other vegetation as it blocks their search for sunlight. Profetto's work also solidifies a long-assumed fact about the vine—it outcompetes other native plant species, a quality that makes encroachment by an invasive species like kudzu threatening to any ecosystem's equilibrium and existing biodiversity. Compounding these impacts, the presence of kudzu invites a myriad of diseases and pests, including the kudzu bug, which induces plant stress by feeding on sap, and Asian soybean rust, a pathogen that threatens soybean production.

In the wide, wondrous world of plants, kudzu resembles a seemingly unstoppable force. The code to its relentless survival is its tap roots—abundant with carbohydrates, the tuberous roots burrow deep into the soil and are able to absorb water from underground, allowing the plant to thrive in hot, dry climates. Using its vines to cut off trees and other plants from the rays of sunshine that sustain them, it overtakes other plants by establishing crowns or new tap roots at nodes along the stems. Another element that sets kudzu apart is its propensity to grow in almost all kinds of soils, persisting in landscapes and terrains where other vegetation cannot survive. It can be found most often in deep, loamy soils. Meanwhile, the emblematic plant spreads in one of three ways: vegetatively produced vines, seed germination, or human intervention. The last is thought to be the most common cause behind kudzu infestations in areas the vine is new to, although kudzu seed spread is highly variable. According to New York Invasive Species Information (NYIS.info) asexual spread, or vegetatively spread kudzu, is more common in the US than sexual spread, or seed germination. Each of its brown, flattened seed pods holds between three and ten seeds.

The vine has been spotted in thirty states and counting, with the latest survey of the USDA's Plants Database mapping the range of kudzu in 2014 to include Hawai'i, Oregon, and Washington. Populations have even been identified as far north as Ontario. Research suggests the vine thrives in climates with at least forty inches of rainfall every year, long growing seasons, and generally milder winters, all staples of the southeastern United States. Climate change is triggering longer growing seasons, more frequent and intensified extreme weather events, and more rainfall—all of which could result in ripe conditions for its spread. Still, much about the "scourge of the South" remains unknown and will likely stay that way unless funding becomes more readily available. "I know that there were other things that I wanted to look at that we just didn't have the funding for," Profetto asserted. "There's way more to look out for kudzu, and actually quantify and figure out, especially as the temperatures keep rising."

Not too long ago, Profetto had hoped to explore what she considers a big gap in our understanding of kudzu—the vine's blossoms. The little flower buds are understudied, she says, and their patterns of growth are unexplained—i.e., it's not clear why the elevated kudzu patches that grow over trees are more prone to have flowers than sheets of kudzu on the ground or wild layers that engulf fields. That's one piece to the puzzle Profetto herself tried to look into. But because existing research on the blossoms was relatively nonexistent, Profetto was unable to pursue her questions any further. "It didn't make sense," she said. "And there has been nothing on it since I tried to do that study . . . it's just so weird."

"Silly questions" like these are stupefying to the scientist. "It's interesting that the plant species that's coined as the 'plant that ate the South,' after all these years, there's no quantitative data. It's baffling that it took this long," she added. "Everybody's fine and they just deal with it. And I think that that's where the lack of research came from. It's just this mentality." The mysteries at the heart of it continue to confound many—with some in the scientific community recently casting skepticism on the vine's fabled relationship with soil stabilization. In a 2016 article published by Sandra Avant, a communications officer at the Agricultural Research Service (the research arm of the US Department of Agriculture), plant pathologist Mark Weaver is quoted as saying that gullies have been found forming under patches of established kudzu—effectively making the case that the weed doesn't control soil erosion, but merely hides degraded land under its sizable growth. In any event, it's worth noting that the bulk of peer-reviewed research to date doesn't support that conclusion.

Now living in Tifton, Georgia, Profetto says she's heard mumblings in the surrounding community that lead her to believe no one really understands what the vine can do—and the majority are okay with that. "Nobody cares about it," she said, noting that this lack of interest is a grievous mistake. "It is still increasing, especially with the increase of climate change. It'll never stop. It'll never go away. Why wouldn't we take the time to look at it? Why wouldn't we take the time to understand it fully and completely? It's just ignorance."

We can't count on Profetto to lead the charge, however. While she "strongly believes" that there is "more research to be done, and should be done," she doubts her ability to continue on with the disheartening battle. Demanding and high-intensity fieldwork is a huge part of why she won't be the one to shoulder the burden. After her multiyear stint, she says she won't be rushing to recreate the experiment anytime soon—if ever again. "I don't think that I have the drive. I'm gonna be honest there," she confessed. What was at first infectious enthusiasm for the vine has been dampened by the physical challenges of the fieldwork.

Just as a stand of ancestral trees encased in kudzu desperately needs the light of day, the world needs more insight into the patterns of this perilous plant. As temperatures intensify and carbon dioxide levels rise, so does the demand for this work—at least according to Profetto. "There's definitely more to unpack," she said. Her five years of research barely scratched the surface. "The fact that we just automatically assume these things about invasive species . . . We really need to look further. There's more to be seen from kudzu."

In the South, especially in Georgia, Alabama, and Mississippi—where, decades ago, the heaviest kudzu infestations were thought to be concentrated—the vine has been characterized as a nuisance by those working in industries battling its path of ecological and economic destruction. Its reputation is shrouded in negativity. Nowadays, kudzu is universally considered aggressively damaging to American biodiversity, economies, and ecosystems. But over the last decade, some communities have also found a new sense of allegiance, and even a begrudgingly positive perception, regarding a vine that's here to stay.

Nestled within the Blue Ridge Mountains of Virginia are five acres housing a vegetable farm and wood-fired bakery. Every spring, over 10,000 seedlings—herbs, flowers, and vegetables—are produced there, while naturally leavened breads and croissants using several ingredients sourced from the land are baked in a 5′ × 7′ brick oven in a 1,000-square-foot building on an expanse of otherworldly greenery. A team of seven runs the show at the Little Hat Creek Farm, five of whom belong to the founding family. Heather Coiner, Ben Stowe, and their trio of children are central to the success of the diversified ecological farm in Roseland, Virginia. After all, it's their brainchild, a business and lifestyle the Coiner-Stowes went all in on in the fall of 2013 after meeting the year before at a music festival. The farm followed on the heels of Coiner's homemade-bread company, Pannier, which she started in Toronto after falling in love with baking while pursuing her doctoral degree in ecology.

An interest in the unknown was what first beckoned her to kudzu. "Oh, it's everywhere here," Coiner said. While the perennial isn't something she deals with on Little Hat Creek Farm's private land, she says it's a staple among the flora and fauna carpeting the mountainous landscape near where she lives. "I see kudzu every day." Coiner isn't one of those who instinctively associate kudzu sightings with negative connotations. She grew up in the deserts of California and was first introduced to the vine when her supervisor at the University of Toronto suggested it as a graduate research topic. "I came in completely as a blank slate, but had already been a little bit, this is strong language, but 'indoctrinated' into the invasive species way of thinking," she explained.

The next six years of her life were devoted to studying the science of kudzu, as Coiner pursued her PhD in plant physiological ecology. Her dissertation focused on exploring unanswered questions about the mystical, symbolism-packed weed. "The thing people focus on is invasive species as embodying this evil. Like a 'Genghis Khan-attack-on-the-landscape,'" she laughed. "But plants will just try to do their best with whatever situation they're put in." It's particularly prescient for kudzu because it is "so maliciously characterized in the popular mindset and also in scientific literature," she noted. By embodying the "invading plant," the ecologist believes that kudzu has been typecast in an antagonistic role as the quintessential "poster child" of invasive species in America. "It always bothered me that when I was working on that, as soon as I mentioned kudzu, people had a cultural understanding of what the plant means and assumptions about why I'm working on it and what I'm finding," she admitted. "Which is different than if I'm working on some non-culturally important plant that doesn't have all that baggage."

The year was 2007. The budding scientist had one singular goal in mind—using kudzu as a means to study the northward migration of plants in the face of worsening climate change. No one else had attempted this before. In fact, Coiner quickly learned that not many had ever looked into kudzu. "I became really frustrated trying to learn more about kudzu because it seemed like every publication was making the same claims about how it doesn't survive in the North without backing it up," she said. It made her, as a young scientist, want to get to the bottom of whether that was true. "It just seemed to all be coming from the same source, or this mythology that was being passed around about kudzu." Undeterred, the researcher took a road trip through North America, collecting samples of numerous kudzu populations, including the first Canadian kudzu population observed, which was discovered in 2009 on a hillside along the north shore of Lake Erie in southern Ontario. That became a field site for Coiner, where she could conduct physiological work in response to actual climate events. She would diligently monitor the weather—when there was a spring frost, she mobilized her research team. Together, they'd head out to the field to try to measure what the kudzu was doing in response to the weather event. It was there, on a south-facing lake shore, in the most temperate area of southern Ontario, that Coiner would track the existing kudzu population and its responses to colder weather for two years, looking at the effects of cold temperature events on the plant, in every season. Although they don't know how long it had been there, there was no sign of other populations elsewhere nearby. She believes it's most likely that the landowner planted it because they had read something about it being good for stabilizing steep slopes.

What Coiner and her team unearthed, in the end, disproved all anecdotal claims that kudzu is unable to handle freezing temperatures. Reviewing fourteen populations of the plant spread across eastern North America, Coiner discovered that the vine survived winters below -20°C (-4°F), meaning that it can persist north of its current range. At its coldest point, temperatures around the plot in Ontario reached -26°C (-14.8°F), suggesting that the plant could successfully thrive in a poleward march—aided especially by our actively warming climate. Research published in 1969 found that the "optimal temperature" for kudzu growth is around 30°C (86°F), which was used to explain why it flourished in growth patterns in the South. In spite of that, Coiner's research reveals that kudzu has the capacity for remarkable cold tolerance, suggesting that as global temperatures have been rising, it has migrated northward—and that low temperatures aren't likely to prevent it from further expansion. "I certainly didn't find cold temperatures to be a silver bullet," she said.

The real question, according to the ecologist, isn't, "Is kudzu going to move northward because of climate change?" but "Are people going to continue to try to move kudzu northward?" "I'd like to shift the focus to the role of humans in dispersing kudzu," Coiner declared. Her work underscores the importance of more stringent management strategies, namely, a collective campaign ensuring that kudzu seeds and plants are not readily available or getting moved around inadvertently by humans. "After doing all this research, the biggest takeaway for me was, maybe it's the wrong question to be asking: 'Are plants moving northward because of anthropogenic climate change?'" The way she looks at it, the question should instead be: "Will plants be able to survive better in more northern areas because of climate change?"

In the case of kudzu, the answer points to *yes,* but it calls for further investigation. Up until this study, it was widely believed and accepted by invasive ecologists that a cold winter was the perfect kryptonite to populations of kudzu—as the weed was said to die back completely due to killing frosts. Coiner's research pinpointed exact instances where that wasn't the case—evidence that kudzu kept thriving despite the weather. What is indisputable is that kudzu does survive in spaces beyond the subtropical ecosystems of the South. "My research shows it's pretty cold tolerant actually, and, if you give it a chance, it will grow," she observed. The thorniest challenge Coiner faced while conducting her probe was a lack of other peer-reviewed research on the plant's propensity for growth. At times she even resorted to using outdated sources like government bulletins to try to fill some glaring gaps about how kudzu impacts the world around it.

In her eyes, Coiner was the only person examining a culturally noteworthy plant—a view that was validated when she suddenly found herself fielding questions about it that expanded well beyond her research hypothesis. "I would get questions like, 'Well, do kudzu's seedlings survive in the wild?' And I wouldn't know, because nobody had studied it," Coiner said. She knows of one unpublished dissertation that examined that question, but no published papers. "I would get questions like, 'How does kudzu disperse?' And I don't know. No one's looked at it," she added. The lack of verified information about kudzu forced her to speculate about some parts of her experiment, which led to numerous publication delays. "It was extremely frustrating for me because I was like, I can't look at *all* of these things," she declared. "I'm one person!" Her conclusion is that it's understudied, and there is so much more to be understood about the weed that "consumed" the South. She isn't the first to call for the investment of more resources into expanding our understanding of the vine—a report on kudzu published in 1985 by Robert Hill at the Pennsylvania Department of Agriculture urged for more research and attention. "A recent library search resulted in virtually no references for biological information or life history studies for kudzu in the northern extension of its range," wrote Hill. "The need to critically investigate this recent addition to our flora is great."

Coiner believes that studying how kudzu is affected by the warming world can help inform ecologists and researchers as they investigate one looming question: How is climate change impacting various species and ecosystems, and what does that mean for humanity? She hopes that computational landscape ecologists and climate scientists eventually end up assimilating all the data on invasive species in a probabilistic model, something that can guide scientists and policymakers as they focus on stabilizing and mitigating the impacts of climate change on rare and endangered native species.

These days, Coiner is no longer analyzing kudzu. Although she's left academia behind in favor of operating Little Hat Creek Farm, if she were ever to return, she says that's exactly where she would pinpoint her efforts. She's optimistic that someone else will eventually build on what she learned investigating kudzu's spread and its connection to our warming planet. But at the moment, she's not sure anyone else is actively pursuing answers to these questions. It's a sentiment shared by entomologist Matthew Frye, who works for the New York State Integrated Pest Management Program at Cornell University. "There's not this vast expanse of literature where people have evaluated it [kudzu] from all different angles. It was, 'Here's the history of it, and here's what's happening with it in its native range, and here's what we've tried to do in the state to control it,'" said Frye. Describing the body of peer-reviewed research as "very limited" in the context of herbicide applications, he expressed surprise that so little effort has been made to untangle the secrets of a weed with a legacy like kudzu.

Not only do Frye and Coiner share a former research focus, but they also happened to be studying kudzu at similar junctures in their graduate trajectories. As a result, their paths inevitably crossed; at one point during her research into its poleward range expansion, Coiner went to harvest roots from an overgrown plot in New York. She removed the root system of one of the plants, while also cutting above the root crown. Lo and behold, that land belonged to Frye's parents—and was the site of a decades-long battle with a verdantly villainized vine. "In undergrad, I never took any botany courses, so I hadn't learned about kudzu or what it was," said Frye. "The first day that I stepped into the lab, I was like 'I know this plant. I have seen this plant my whole life.'" Since 1974, the year his parents bought their home, his dad, Francis, had been trying to curb an overabundance of kudzu entrenched on their family's property. When he was a kid, the perennial was a source of endless entertainment for him and his siblings. Frye remembers afternoons lost to adventures in the backyard, aided by stockpiles of the towering greenery. "There were these kudzu vines hanging from trees in our yard," he said, recalling how they would play with them for hours on end. It wasn't until that day in grad school that he knew exactly what that hallmark of his adolescence was. "That was another full-circle moment for me. Realizing that this vine that my brothers and I had done Tarzan swings on across the yard was actually kudzu."

It was 2004 when Frye dove full-speed into studying kudzu at the University of Delaware. The lab's main project was studying mile-a-minute weed, but his isolated focus was on the monstrous foe encasing southern landscapes. He found it curious that most of his peers had never heard of the "vine that ate the South." "I always felt like the redheaded stepchild in that case, because I'd be like 'Oh, I study kudzu.' And everybody was like, 'What?'" All of his investigations into the perennial had to do with biological control methods—asking questions like: "What type of damage to the plant would have the greatest impact on its growth and reproduction? What type of insect should we focus on finding in the native range, that would have the greatest impact on the plant's growth and reproduction?"

Kudzu is eradicated in a variety of ways, and researchers are divided on what the best management strategies are. When it comes to controlling it, patience is a prerequisite. "Our results, and the work of others, have demonstrated that it does take time," Frye said. Applying herbicides to expansive areas overtaken by the plant usually leads to a lengthy waiting game. The USDA Agricultural Research Service (ARS) surmises that controlling kudzu stands can take as long as a decade, with "persistent herbicide applications" usually needed to get rid of it, although a 2016 ARS study found a heady cocktail of four herbicides mixed with a bioherbicide treatment was effective at nearly eradicating sites of kudzu within two years. Herbicides are a widely deployed option by the Department of Transportation in several southern states, but some invasive species specialists tend to shy away from this recommendation due to the harmful ecological implications—the global farming industry's reliance on the weed killer Roundup and its ensuing effect on biodiversity is perhaps the most infamous of cases—as well as the muted effectiveness on herbicide-resistant weeds. A whitepaper published by the Virginia Department of Conservation and Recreation and the Virginia Native Plant Society suggests that the most effective method of control depends on the size and accessibility of the patch, as well as its proximity to other flora and fauna. Removing the roots in a kudzu stand is a universally acknowledged path to success—but it must be done with persistence. With root systems that dive belowground, sometimes embedding twelve feet under, killing kudzu by direct root removal can often be impossible.

For six years, Frye experimented with biological control management procedures for kudzu. To conduct that research, Frye grew 105 kudzu plants in a greenhouse from seed, which were later transported to a field owned by the University of Delaware. From there, he applied four different known management techniques to test which control strategy was most effective. His purpose was to determine what type of insects would have a higher impact on the plant—leaf-eating vs. tip-damaging or root-damaging bugs. Frye describes it as "feasibility for insect biocontrol," but looking at the different types of damage on plants. The methods tested included cutting up to three-quarters of each kudzu leaf and shearing down shoot tips to the same degree, and then allowing 35 plants to grow for one growing season, and another 35 to grow for two.

The result was just as Frye and his associates predicted—heavy defoliation of the plant did appear to damage the kudzu, suppressing growth and producing shorter vines and less biomass. Frye's takeaway is that repeated defoliation, or cutting down the leaves of each plant, is another effective control strategy. "The research actually demonstrated that removing lots of leaf material, so up to 75% of each leaf, had an impact on how the plant grows," observed Frye. "In terms of its architecture, everything was shorter. The vines were shorter, and it just made a more compact and stressed-out plant." It also provided a case for defoliation as an effective method—but one that, like the rest, takes a long time to see to fruition. "The benefit of using biological control is that you're investing time, but it's with insects, and so they're the ones doing the damage. It doesn't incur the cost of spraying several times a year, over multiple years," he noted.

To date, there hasn't been a single insect identified that could act as an effective biological control agent for kudzu. But if you ask Frye, his work not only backed up existing work in the research field but also provided a different lens to expand what we know about the vine. "It was a different approach to understanding how this plant is growing and how it uses resources to take over different environments because it outcompetes other plants for light," he said. Although defoliation proved to be effective, Frye says shearing kudzu leaves is not the "default" kudzu control. One of the challenges in the management of an invasive plant is resource investment. According to the entomologist, the easiest way to manage large-scale plant populations is through herbicide applications. "The root crown removal is extremely labor intensive. That's why people aren't really implementing this on a large scale," explained Frye. It would require effectively cutting out the root crown of a kudzu plant, making sure that you've removed all the root crowns in an area, and then, in the case of large-scale invasive plant removals, revegetating the area with a native species that belongs there. Not all roots need to be removed to kill the plant—just the crown or the portion at the interface of the below-ground roots and the above-ground parts of the plants. "And for some of the infestations in the South, I'm sure it seems like it's unfeasible to do," he said. If that final management step isn't taken, he suggests that the outcome would likely be dire—as it could intensify the likelihood of another invasive plant potentially invading the zone, taking over kudzu's former playground. Like extending an invitation to other sprawling beings on an instinctual hunt for new places to claim as their own. "It's called the 'invasive species treadmill,'" Frye said. "Where you remove one, and then the next one comes in, and you remove that and the next one comes in. So, without any intentionality [in] filling that void, you're going to have to continue to manage that site." He urges anyone looking to remove kudzu from their property to consider the next step in the whole management program.

A few summers after Coiner took samples of kudzu from a plot teeming with it in New York, Frye and his dad would replicate the removal of the root systems from bands of the plant on his parents' property. Years later, they noticed their manual procedure worked. In 2009, thirty-five years after Frye's father began his battle with the persistent vine, he finally laid down his arms, as the patches of kudzu retreated once and for all. "We were shocked to find that now there's no kudzu where they live," he said. "After however many years, cutting back the vine, my dad finally won."

Thanks to the time he spent researching the weed, Frye carries a wealth of information about the "vine that ate the South," and with it, a weight of responsibility. The entomologist is hyper-aware of the ecological consequences of invasive infestations, and the detrimental impacts that could domino out if misinformed people began cultivating kudzu. For the last decade, as one of a handful of researchers who has studied the perennial, he has fielded requests from people across the country who are interested in harvesting it. The purposes vary. An artist working on an exhibit wanted to grow kudzu to have it draping and growing all over their display (Frye said "yes" *only* after he worked out a plan with the artist to dispose of it after), a researcher from Michigan wanted to use parts of kudzu's genome in other plants as a way to end world hunger (Frye decidedly did not entertain that one), and a forager reached out in an attempt to plant it somewhere in the Catskill Mountains, a terrain with abundant native species (another case where the entomologist said "absolutely not" and added, "The last thing I would want to be responsible for is having kudzu planted there"). More recently, he has had a student in Kentucky inquire about how to grow it for a science fair project on the allelopathic[2] effect of native plants on kudzu seeds (which they won a prize for), and some researchers in South Africa wanted his help with their risk analysis of the vine as an "alien species." "There's this constant barrage of solicitation of information about kudzu. Even though I'm in a totally different field, I can't seem to escape it," expressed Frye, who finds it baffling that, all these years later, he's still on the receiving end of regular questions related to the vine. "Because kudzu is always coming back."

One striking request came from an elementary school class in Alabama that was interested in having kudzu seeds sent into space to determine if they would germinate under zero gravity. They were working with a NASA-affiliated educational program. "Kudzu in outer space sounds like a horror movie," Frye said, chuckling. It could be pulled straight from a post-apocalyptic sci-fi script—except, in this case, it's real life.

2. Allelopathy is defined by the *Encyclopedia of Applied Plant Sciences* (Netherlands: Elsevier Science, 2016) as "a direct or indirect interaction, whereby allelochemicals released by one organism influence the physiological processes of other environmental factor neighboring organisms."

4

BEAUTY FROM THE BEAST

Five years ago, in the middle of a sweltering September day in Winfield, Alabama, a seventh grader sat in a science classroom thinking hard about a vine that had inadvertently been a staple of his childhood. Will May was brainstorming with a handful of his peers, going over possible topics for their upcoming group project. Their assignment was straightforward: Their teacher had asked the class to come up with something they could grow in zero gravity.

The idea to use kudzu in this space project materialized thanks to an idle suggestion from one of the other students. "Someone looked out the [classroom] window and was like, 'Well, why don't we look into kudzu?'" A glance outside of that very same window, a portal into a world bursting with flora and fauna, would reveal a wall of Winfield Middle School partially swallowed by the "vine that swallowed the South." Gaze locked on the sheet of kudzu gift-wrapping the building he sat in, May couldn't think of a reason why they shouldn't. The case to use kudzu in their space-based reconnaissance was met with unanimous approval; the jury was a group of twelve- and thirteen-year-olds eager to get on with their assignment. "We were like, 'Well, might as well,'" May said, remembering the collective muted enthusiasm of his group of middle-schoolers. With time, that zest would grow, much like the brightly hued blossoms that tend to adorn hanging kudzu vines in the fall. As they read up on the weed's background and learned about its rampant growth, versatility in cooking, and nutritional benefits—the crew of seventh graders "couldn't believe it" when they also came across research suggesting that the fabled weed could be a source of antioxidants—they explored how it could hypothetically be put to use for future communities. It could possibly even be used to help the survival of their descendants, who could one day live among the stars above.

May's class at Winfield Middle School was taught by Freda Curd, a now-retired science teacher with a background in education spanning nearly three decades. Curd instructed the seventh graders in her charge to write proposals for the Student Spaceflight Experiments Program, a national student competition operated out of the National Center for Earth and Space Science Education, that allows students to design research for NASA and the International Space Station (ISS). The teacher thought May's project—and the group's reasons for choosing to send kudzu to space—was one of a kind. It was, according to Curd, an inimitable plan, given that the group of seventh graders ended up uncovering a host of benefits associated with the vine. "It's not if we go to Mars, it's when we go to Mars, and we're going to have to find ways to feed people and keep people alive," Curd explained. It wasn't long before she became one of several people impressed by the novelty of this idea. Much to their surprise, the proposal ended up being selected for the next level of the competition. May and his fellow "space kudzu pioneers"—Seth Birdsong, Cole Kirkpatrick, Banks Roebuck, and Izzy Stewart—got to test their written theory, first proving in a series of experiments that they could germinate kudzu seeds here on Earth. And before they knew it, they were invited to a national conference with the other participating groups. It was there, at the Smithsonian National Air and Space Museum in Washington, DC, that a group of seventh graders from Alabama got to present their science project. The crowd was made up of other participants, students, educators, and judges—which included an array of people from across the country who had never heard of the "vine that ate the South." "We got to educate a lot of people on kudzu," asserted May.

Following that presentation, they moved forward to the final stage—shipping the invasive plant's seeds to the galaxies above. This wasn't the first time that NASA had dabbled with kudzu—but the first time it entertained sending it to space. In 1979, NASA scientists B. C. "Bill" Wolverton and Rebecca C. McDonald published a report that summarized the results of an investigation experimenting with anaerobically fermenting[1] the plant biomass of kudzu, water hyacinth, duckweed, and water pennywort as sources of renewable energy. The two wrote that the work came as a response to the US

---

1. The 2011 *Encyclopedia of Dairy Sciences* describes fermentation as a "partial metabolizing of nutrients consumed by the host, providing energy and nutrients needed by the microbes" (https://doi.org/10.1016/b978-0-12-374407-4.00371-x). In other words, it's a process that allows for the production of energy without oxygen.

Department of Energy's consideration of "large-scale energy farms" that cultivate plants "solely for the purpose of fuel production." Kudzu was chosen as an "ideal candidate for energy farms," due to its ability to grow quickly, its long roots that allow it to "withstand droughts," its strength, as measured by its "resistance to eradication," and its ability to "thrive in poor soil that is useless for agriculture and actually improves the land by restoring nitrogen to the soil." The results of that research demonstrated that anaerobic filters reduced the total digestion time from 90 days to 23 days on average—and found that kudzu had a "high potential methane production per unit dry mass." The report ends by recommending kudzu as an "ideal candidate for terrestrial energy farms on land that is not suitable for agricultural use."

Nearly four decades later, a group of seventh graders have taken the idea of kudzu as a renewable resource to another level. In order to fully test their experiment, they needed to ship eight kudzu seeds to be germinated in space, which they sent along with distilled water and formalin, items that, if carefully administered and monitored, would preserve the ingredients once germinated. Altogether, those ingredients would travel more than two hundred nautical miles above the planet. Their destination? The ISS itself. There, scientists would use the students' proposal as a guide to help test their theory: Could the "plant that consumed the South" germinate in zero gravity?

Not long after, their winning project was sent to the world beyond this one. And almost a year after they began, the class at Winfield Middle School got a package in the mail. It was the results of the experiment they had sent to NASA. "It was like opening the best Christmas present," said Curd. The group raced to their classroom lab to finalize the results of their work—they would determine whether they were successful by running experiments at the ground level and at the ISS station, comparing the kudzu seeds' stem lengths and biomass to measure the effect of gravity on growth. If the package from space showed the same or a greater percentage of germination, length of seed stems, and amount of biomass, then they would have quantitative evidence of what they hoped would be true—even a lack of gravity can't deter kudzu. And that's exactly what they discovered. Against all odds, their project was a success. The kudzu seeds, prepared by a group of middle-schoolers, had grown in outer space. Not only that, but these experiments revealed something nearly impossible to grasp—the seeds had positively flourished in microgravity. "Kudzu grew *better* in space than it did on Earth," May pronounced incredulously.

Born and raised in Alabama, May has a lifetime of familiarity with the arresting weed at the center of their pioneering school project. It had slithered and coiled its way across the fields where he was raised and was universally considered a nuisance to those in his community working in agriculture—like May's own grandad, who was a farmer. "We would go out and see it everywhere around the farm. It had just taken over trees. There was a barn he had that was completely covered in kudzu," May remembered. "I've always been around it and I've known what it was." But it wasn't until that middle school class where he started experimenting with kudzu that he first learned about the endless ways it is being used in other countries. "I never knew you could *eat* it!" he exclaimed. Now that he's more informed about the utility of the vine, and its varied history of international applications, he laments how it is perceived by those around him. "It shouldn't be as villainized as it is," he said. Experimenting with weeds in space helped solidify an avid interest in the world among the stars for the then-thirteen-year-old. Now a high-school senior, May dreams about being an aerospace engineer for NASA—where he one day could very well continue pioneering experiments with a plant so many of his neighbors consider a plague.

His seventh-grade teacher would love to see that destiny unfold for one of her former star students. Enlightening and educational, Curd's role in guiding the teens with their kudzu experiment offered her something just as valuable—it amplified an existing appreciation for a leafy emblem of the South. Curd grew up two hundred yards from where she lives now, in Hackleburg, Alabama, a rural town a little over thirty miles north of Winfield. "As a kid, I just didn't really think that much about kudzu," she admitted. "I knew what it was, and I knew it grew really fast and covered everything." Yet as the landscape around her has changed, so have her opinions of the climbing vine. Today, it's no longer the forgettable fixture it once was. Instead of willful ignorance, or an instinctual irritation, what she feels for kudzu is a lot like gratitude, mixed with a sprinkling of awe.

The credit doesn't solely belong to the seeds they sent to space. More than a decade ago, the former science teacher's hometown was hit by a tornado—one so devastating that even now, the community is still working to rebuild to what they were before it. On April 27, 2011, an EF-5 tornado barreled through Hackleburg, on a path of destruction across Marion County. Bearing winds up to 210 miles per hour, it was three-quarters of a mile wide, descending with little mercy upon the Alabama town of about 1,500 residents. Once the catastrophic swirling column of wind dissipated, 100 people were injured, 18 people were killed, and about two-thirds of all of the buildings and houses in the area were decimated. Everything was gone—including some of the lush vegetation that had once flourished across Hackleburg's fifteen square miles of countryside. Where ethereal trees had once reigned supreme lay row upon row of razed fields, uprooted vegetation, and barren roads. "And that's just a devastating look. When there are no trees, no anything," she recalled.

A year after the devastation struck, amid the wreckage, a startling sign of new life appeared: A strip of shimmering vines emerged from the ground, an unlikely symbol of resilience. "The kudzu came in. And it's covering things, and it almost made it bearable," Curd said. She believes kudzu's ability to flourish in the aftermath, growing despite the tornado's mighty wrath upon the lands of her cherished hometown, didn't just help fill the ravaged landscape— its presence helped community members, like herself, to heal.

"It's almost like a piece of art," she said.

A little over three hundred miles northeast of Hackleburg is Walhalla, South Carolina, a historically Indigenous town with more than four thousand residents. It's there, deep in the majestic foothills of the Blue Ridge Mountains, in a stately one-hundred-year-old house, that Nancy Basket crafts with kudzu. Dwelling on almost an acre of land, sheets of the vine flock her antique property, like crows holding court in a thrush of trees, taking over about a third of the land that belongs to Basket. "It greets people when they come in," Basket said. When she found the antiquated property, which she says she saved from being condemned, the weed was nearly swallowing the structure. "Kudzu was growing in the front door," she noted. "I said, 'Honey, you're the one for me.'"

Simply put, Basket is unreservedly obsessed with kudzu, an adoration that's reflected in her incorporation of the nefarious plant into nearly every aspect of her day-to-day life. Sheets of draping vines, which Basket has made a livelihood from, surround her home, riotously bursting out of a steep, adjacent ravine, where some of her neighbors have "poisoned" the towering weeds they find encroaching on their properties. Not Basket. "I want to keep the kudzu behind my house," she confirmed, a confident smile spreading across her lined, animated face. "I have loved kudzu for the thirty years that I've been in this state." A seventy-year-old artist and educator, she is the founder of Kudzu Kabin Designs, where she crafts and sells everything from chandeliers to miniature sculptures to baskets constructed from the plant in question. Cloth made from parts of the perennial is another of the artist's favored uses of kudzu—as is paper. She tries to use everything she collects from kudzu patches for something, intent to never allow portions of it to go to waste. Whether it's for the products she sells, the food she makes, or the classes she teaches at local schools, Basket believes the vine offers many uses that most struggle to see. "I like doing everything I can with it," she said. Come collection time, she'll venture into the ravine behind her house or head out to sites nearby where she's identified masses of the plant—like plots of regal oak trees wearing kudzu like a close-knit sweater. While she's collecting kudzu to create with, the artist makes sure to honor the natural world she ventures into. Basket is Indigenous—her dad is Cherokee, while her mom is German. Centric to Cherokee tradition is a call to respect the earth and its components. "Everything has a spirit. Everything is living," she explained. "We talk to all of the beings: 'Tell me how you want to be used.'"

In the thirty years she's spent knee-deep in ample thickets of vegetation, gathering kudzu, Basket has never once come across a snake. It's a point of pride for her—and the result of a regular communique with the slinky, nimble creatures that live in the foliage. "You're going into somebody else's home. All beings are sentient. They have as much right to be as I do," she noted. By paying respect to the other beings that dwell there, Basket shares her secret to gathering kudzu's roots and leaves—a chant, always paired with a dance. "I say 'Brother, Sister Snake, I'm coming through! Watch out for me, I'm watching out for you!'" Every trip she makes finds her tapping into her connection with the plant, as strong and indestructible as an old, expansive plot of tuberous tap roots that's settled down into the ravenous soils of the South. After all, that unspoken bond between woman and vine is how she knows the best ways to collect and create with kudzu. "[The vines] are my friends," she declared. "They helped me make a living for my family, for all this time. They grow twelve inches a day, working out, trying to find somebody who loves them. That's me."

Basket's bond with kudzu began in 1989 when she moved to South Carolina. Not long after her arrival in Walhalla, she happened upon news of a local kudzu festival. Already a seasoned basket-maker—the artist says she'd been crafting the object for about a decade by then—Basket found herself immediately intrigued, especially after hearing about the locals that would drag the plant down the street as they mocked it for existing. Outspoken derision like that, leveled at something that sprang from the land, roused the opposite feeling in Basket—she decided to try to build with the target of everyone else's disdain. Her first attempt at making a basket out of the vine led to a flimsy construct—within four days, the item had fallen apart. At the time, she had just started learning Cherokee stories, embracing more of her Indigenous ancestry. The symbolic breaking down of the kudzu-based basket was a sign that could not be ignored. In Cherokee culture, protecting the natural world is sacred. All elements of nature, from the moon to the plants, carry a spirit. Traditional ecological knowledge is different for every tribe, but all share a core responsibility to preserve natural resources, to stand for Mother Earth. Native people worldwide have embraced the environment and fought to preserve it for thousands of years. Not just for themselves, or for their children or grandchildren, but for the next seven generations to come. Storytelling plays a big part in that—passing on legends and finding new ones in the world around them. It's just a matter of listening—with your head and with your heart. These are lessons Basket knew and held dear.

"I thought, 'Oh my gosh, if I'm not going to put the stories into action and into my life, I'm not going to get any more stories,'" Basket said, noting how she went back to the kudzu and addressed it directly: "I'm sorry," Basket told the vine. "I'm northern. I go too fast. Please tell me how you want to be used." The towering patch of kudzu that's long been a staple of her life became a messenger, using the sounds of nature and its enchantment to answer her question. "And they said, 'Leave the trees alone,'" recalled Basket. Trees are entities worthy of respect to the Cherokee—cedar, holly, pine, spruce, and laurel in particular. They are considered powerful hosts, instrumental in the creation of the planet. "Use us for paper, instead."

So she did. Paper made from kudzu leaves is now one of Basket's favorite things to forge—and incidentally one of her best-selling products at Kudzu Kabin Designs, a one-woman entrepreneurial operation run out of her century-old house. Thanks to a dash of ingenuity and loads of tireless experimentation, she taught herself how to split a kudzu root—or, rather, the vines themselves shared that knowledge with her. "You don't just quit," she declared, talking about the many trials and errors encountered while getting familiar with new material. "I was *given* how to split the vines." Nowadays, she has to drive around to find other parcels of kudzu, as she's done so much collecting of the weed in her backyard oasis and prefers to use kudzu that grows in trees. Those hanging vines tend to be thicker, which comes in handy when crafting with it—especially when creating baskets. The artist also likes collecting newer plots of kudzu, green vines still in their first year of growth, to apply to some of her creations. In order to make her namesake, Basket cuts the vines into three-foot sections and then splits them down the middle. Whether she's incorporating its elements like blossoms into a beverage or creating paper from its leaves, no part of the kudzu she collects ends up discarded. "I use every single piece," she said.

Five years ago, Basket traveled to Japan, where the plant originated. Surrounded by a countryside brimming with cherry blossoms was where she met with people whose ancestors have long celebrated it, eagerly offering anecdotes about the generations of their family members who have made cloth out of the vine, a practice they still honor today. "It gives me goosebumps now," she said, remembering how they took her to "meet" their kudzu patch. "They introduced me to this patch, and that is what has fed their family for seventy years," she said. Her relationship with kudzu as a source of her livelihood is one the artist wants people to know is full of respect. "I'm not killing kudzu. I'm not using poison. I let it go," she noted. "You're not hurting a plant. They're giving you permission to do this because they branch out. I'm not eradicating it. I'm just using it in a way that is pleasing to both beings."

Because of her work and overall influence in the American kudzu community, people have tried to give her the nickname of "Kudzu Queen." "I reject that," she announced. "Native people don't have queens." Someone else once told her she was the kudzu *huaca,* which exists in modern-day cultures but translates to "spirits" or "sacredness" in the Inca religion. Basket likes that suggestion best.

If you ask Basket, the plant first caught her attention many decades ago because of its remarkable physical aesthetic. An aura of radiant energy positively thrums within a strand of kudzu, according to the Indigenous artist. It's alive. And in each shoot and stem, as intricate and important as its DNA, is a lesson. "There's a mystique about kudzu. There are legends about kudzu," she said. "We've got to find out what the real legends are and the truth behind them." The vine being a "scourge of the South" doesn't fall under the latter. Labeling plants as "invasive," "non-native," or "native" is not a practice that Indigenous traditions like the Cherokee subscribe to. As far as Basket is concerned, nothing good comes out of that distinction. It's a system that allows humans unlimited reign over animals and plants and lands that do not belong to us. "Native people have always been here, and our stories go along with telling you how to use what you have . . . [and] be one with how we can help each other," she described. "Everything was made here on the earth before we got here. It's our job to be human and find out how to live with everything else that we've been given. So when you think that you've been given a raw deal, because you have nothing but kudzu, well think again. Find out how other people are doing it, and do that. Steal my ideas, please."

Some invasive species researchers give pause to this idea and the capitalistic inferno it could spark. If too many communities were to get on board with repurposing kudzu in ways that may lead to mass production, they warn it could quickly become a complication that fuels a whole other set of ecological issues. Based in Alabama, Nancy Loewenstein is one of those who is wary; she's an extension specialist in Forestry, Wildlife and Natural Resources at Auburn University, with a focus on invasive plants. For more than fifteen years, Loewenstein has worked in the field of plant extension outreach, which includes multiple collaborations on revisions to an Alabama Cooperative Extension Service article on the history and uses of kudzu in the Southeast, originally published in 1990 by James Miller, John Everest, Donald Ball, and Michael Patterson. "Kudzu is everywhere, and most people recognize it. It serves to some degree as a poster child, because you can just say 'kudzu' and eyeballs roll. People know what you're talking about," Loewenstein said.

The revised article includes references to kudzu's propensity to smother seedlings and saplings and increase windthrow threats for larger trees. It also mentions how utility and railroad companies battle kudzu overgrowth, investing "significant amounts of time and money controlling kudzu growing over utility poles and railroad tracks," as well as federal and state agencies trying to control the plant when it threatens national monuments or historic sites. While some cases of communities and individuals repurposing kudzu are positive steps in the name of mitigating an ecologically harmful plant, Loewenstein warns of the dangers of relying on an invasive plant as a resource, especially in an economic sense. "It's such a cautionary tale, that we don't learn from," she noted. What the invasive species specialist wants people to understand is that kudzu represents an age-old story, one that begins with hints of a happy ending but quickly morphs into a nightmare. A drama where humans bring about their own demise—and it's one that seems stuck on repeat. "We repeat these kinds of things all through society, right? I don't want to get philosophical, but it's like, this seemed like an answer to prayers: 'We're going to fix this erosion problem,'" cited Loewenstein, pointing to the frantic planting of kudzu seeds in the twentieth century and all of the unintended consequences. "We keep doing it with so many of the plants we bring in with great promises, without really thinking through potential outcomes."

But if you ask proponents like Basket, her long-term goal isn't to give birth to a capitalistic industry steeped in cash windfalls from the vine. Her aims are a lot more low-key, and they involve making paper and art and teaching others how to embrace kudzu. "Am I out to change the world? No. That's too big a job," she proclaimed. "Can I affect the lives of the kids and the people that are drawn here? Yes. That's why I am open for business." Even at seventy, Basket has no plans to slow down her work with kudzu. The parables she tells with it are meant to be carried on—stories where humans revere the earth and its gifts. After all, the vine is still wrapped around the outside of her home, rows of large sheets of arresting leaves embracing the columns of her front porch, both spilling inside and trailing out to the waiting world beyond her doorstep.

"I'll work on it 'til my last breath."

5

FOREST
TO FORK

Beads of sweat drip down my overheated body as I ruminate on why I'm pursuing this chronicle of a weed that thrives in disturbed spaces. For what other reason would I be standing in the thick of a stand of kudzu-consumed yellow-birch trees, on the fringes of the Appalachian Mountains in the dead of summer, feeling my sanity slip away with every desperate gulp of water, on the verge of consuming the very object that has haunted my every waking moment over the last year? I tentatively take the kudzu blossom offered to me, assessing it for a split second as its crinkled edges soften in the fading afternoon light, a perfect case study of the art of persuasion.

My eyes drink in every little crease and groove, admiring its indistinguishable color, a melting pot of dark reds and midnight blues and plum purples. A flowery, wild-berry aroma hangs heavy in the air around me, as I furrow my nose in response to the blossom's unexpectedly sweet smell, so strong it could be synthetic. A bed of fuzzy hairs covers the accompanying oval-shaped leaflet, which is more than slightly off-putting. For a moment, I'm unsure if I want to proceed. "What if I hate it?" I ask myself, visions of my body abruptly rejecting this plant, something that's been consumed by people for a millennium, crowding my mind. "What then?" Perhaps this ode I've invested in, to a plant and the places connected by it, will be entirely without purpose. Perhaps whether I like it doesn't matter at all. A few seconds pass that might as well be hours, moments in time where an internal battle reaches its peak: To try, or not to try. Decision made, my hand slips the paper-thin meal onto my tongue, where my tastebuds immediately rise up to meet an almost curiously bland flavor, blanketed in an aftertaste reminiscent of sweetened spinach. I don't love it, but it's also not the worst thing I've ever consumed. It's bitter and it's bizarre—but in so many ways, so am I. Based on its texture and taste, I'm intrigued. I wonder—"Could this be a welcome addition to an adventurous, eco-friendly kitchen?"

Deep fried, sauteed, flavored, or stirred, kudzu has been incorporated into the East Asian culinary craft for thousands of years. Starches made from powdering the plant's root have ancient origins in Chinese and Japanese cuisine. But in the US, cooking with the weed is more of a recent phenomenon. Spurred by the modern revival of foraging and a renewed social focus on environmental preservation—emerging from an era of global pandemics and simultaneous climate disasters—esteemed chefs and everyday people alike have begun working invasive species like kudzu into their recipes. With the exception of its seeds and seed pods, the vine that "feasted" on the South is almost entirely edible. Its tips, blossoms, roots, and leaves can be served up for safe eating. To this day, researchers at the Alabama Cooperative Extension System suggest that flour made from kudzu is imported to the US and can be found in "many Asian grocery and health food stores." The blossoms tend to be purple-colored and smell like grapes; the roots have been compared to a potato-like taste and consistency. The latter can account for up to 40% of kudzu's plant biomass, and the nutritional and medicinal properties of the root have led to nicknames like "longevity powder" or "Asian ginseng." In a 2009 *Kitchn* article, Kathryn Hill succinctly captures the uncapped potential of kudzu in the kitchen: "The leaves can be used like spinach and eaten raw, chopped up and baked in quiches, cooked like collards, or deep fried," Hill wrote. "Young kudzu shoots are tender and taste similar to snow peas."

It may sound bizarre to some, but invasive-based dishes can be found in a handful of restaurants scattered across the Southeast. From Georgia to Tennessee, flowering kudzu has been incorporated into everything from pie to wine. One of those culinary connoisseurs making dessert delicacies with it is award-winning chef José Gutierrez. Published in *Eat the Invaders,* Gutierrez's kudzu sorbet recipe involves whisking kudzu blossoms with white wine, licorice root, cayenne pepper, sugar, and water. Another source of creativity is Angela Gillaspie, who goes by the self-proclaimed moniker "Queen of Kudzu." Gillaspie is the voice powering SouthernAngel.com; part blog, part DIY-source, and part cookbook, the website shares recipes for rolled or deep-fried kudzu leaves, kudzu tea, kudzu jelly, and kudzu quiche. Meanwhile, an assortment of recipes incorporating the weed can be found on Food.com, a hub of instructions on how to make everything from sweet kudzu blossom jelly to a flavorful pork tenderloin with chilled kudzu salsa on the side.

The vine is also popular in liquid form. Kudzu root–based teas have long been common in China and Japan, promoted in everything from weight loss supplements to drinks for an upset stomach. In Georgia, one chef incorporates kudzu into a beloved quintessentially southern beverage—lemonade. Blending kudzu flower syrup, lemon juice, and gin, Mimi Maumus serves up fresh kudzu lemonade at *home.made*, her restaurant and catering company. Located in Athens, home.made has served the refreshing Asian-southern fusion beverage as a menu staple for at least five years. A wild plant enthusiast and professional chef, Maumus was first taken in by the smell of its blossoms. "I lived on this little street. And there was kudzu growing right across the street from me, and I didn't even realize it was kudzu because it wasn't taking anything over," said Maumus. She remembers it being high up, a tangle of vines draped along power lines. One day, she found herself entranced by the viridescent mass towering in the sky. "I had paused and I was just standing in front of it, and I had a neighbor who was walking her dog, [who said] 'Smell that grape Kool-Aid?'" "I *do*," Maumus remembered saying with inflection—to which her neighbor told her matter-of-factly, "Yep, that's the kudzu." Maumus loosely knew it in the way many from the South do—a nuisance of a plant that crowded roadways and killed crops across the Peach State.[1] At the time, the chef wasn't aware that the climbing vine that had caught her gaze was edible. After her neighbor pointed out the blossoms to her, the source of the distinct aroma, she vividly remembers thinking: "'Wow . . . it does have this almost artificial grape smell.'" For a moment, she noted how interesting a plant it was, but still never once thought, "'Let me try and eat some.'"

A love affair with experimenting with wild edibles in her dishes would lead Maumus right back to the striking vine that she couldn't forget—that memory of its reigning stance, clinging to points in the sky cemented in her mind.

1. Georgia's nickname, "the Peach State," was rooted in white supremacy. This is explored in William Thomas Okie's book *The Georgia Peach* and covered in a 2017 NPR article by Tove Danovich: "The Un-Pretty History of Georgia's Iconic Peach," https://www.npr.org/sections/thesalt/2017/07/21/537926947/the-un-pretty-history-of-georgias-iconic-peach.

In 2015, Maumus went foraging with the owners of Bartram Trail Farm in Winterville, Georgia—a local source of seasonal organic vegetables for chefs just a fifteen-minute drive outside of Athens. "There are so many wild edibles that most people don't know about," she said. Maumus's journey with kudzu weaves into her relationship with incorporating wild edibles into her culinary craft. "I fantasize about them, I romanticize them," she proclaimed. By then, the chef was familiar with the poster child of invasive species and had already begun trials with it in her frying pan. The farm was run by Jason Jones and Scott Brandis, who helped her in sourcing kudzu leaves for those kitchen-table tests. "They did have kudzu on their property, and so they went through and they were like, 'We found this,' and 'This leaf isn't good,'" recalled Maumus. She would learn from the farmers that the types of leaves used in her dishes mattered, and through her personal experiments she also discovered that the ideal parts of kudzu to cook with were the baby leaves of a petite variety—not quite as fuzzy, or as fibrous as the rest. "I could cope with those," she noted. It wasn't long before she stumbled upon the best way to serve kudzu to her customers: to get rid of the plant's natural layer of fuzz, which could be done by frying the leaves.

Spellbound by that powerful, "purple Kool-Aid" blossom scent, Maumus was motivated to make a simple syrup out of kudzu flowers, with the idea to turn it into a lemonade. That fall, she gathered a bunch of the plant's flowers and whipped up a recipe. "I was imagining it being this beautiful fuchsia color, the way the flowers are themselves," said Maumus. But when she first made the simple syrup, it surprised her—instead of a beautiful, bright shade, the mixture ended up a gray, dirty-water color. "I was like, 'What a disappointment,'" she recalled. But not all was lost. Although the color wasn't appealing, the chef found the taste of the syrup to be "delightful." Channeling her inner chemist, Maumus experimented with making the result more pleasing, landing on a solution where, if acidulated, the dull color of the mixture vanished and the vibrant splash of the blossoms came out in full force. "It became this magic trick that I started doing around the kitchen," she said. "I felt like Willy Wonka. Like I had just found some crazy new thing."

Kudzu's got a little more than enchantment to its appeal—the plant also packs plenty of nutritional power. In 2014, a report published in the *Journal of the Alabama Academy of Science* compared the antioxidant levels of kudzu to levels belonging to spinach and shiitake mushrooms. Authors Safaa H. Al-Hamdani and David Marc Ponder found that kudzu had measurably higher levels of antioxidants than its counterparts—a benefit that compounds its advantages as a source of plant protein, which was discovered in a 1983 analysis by James A. Duke, a botanist for the Department of Agriculture. Duke found that per 100 grams, raw roots and cooked kudzu leaves offer 2.1 and 0.4 grams of protein respectively. Kudzu was one of many plants Duke would go on to recommend for medicinal treatment application, in books like his 1997 title, *The Green Pharmacy: New Discoveries in Herbal Remedies for Common Diseases and Conditions from the World's Foremost Authority on Healing Herbs.*

A prospective nutritional boost isn't the major draw behind Chef Maumus's kudzu-based concoction. It's all about the presentation. Once she landed on a recipe for kudzu lemonade she loved, Maumus started taking the ingredients for her newly created beverage to others working in her restaurant. Made from a pint of sugar, some water, and two pints of kudzu blossoms, she'd give them a full shot of already prepared simple syrup and just half of a lemon before instructing the recipient to add lemon juice to the thick mixture, stir it, and then sit back and observe what comes next. Without fail, everyone exclaimed elatedly as they witnessed how merging the ingredients would immediately turn the liquid from a dull slate to a gorgeous fuchsia tint. "It was just blowing people's minds and it was *delicious*," she declared. It wasn't long before Maumus added kudzu lemonade to the restaurant's permanent drink menu—eventually even serving it with added flair.

That level of performative panache is now a given. If you order the lemonade at home.made today, chances are you'll get it presented on a wooden platter, with different colored liquids laid out in scientific beakers, all materials evoking a Frankenstein-esque sense of curiosity and the thrill of adventure. From there, it's up to the customer to mix the liquids into a glass, so they can experience the color transformation firsthand. "It was like this mad scientist moment," Maumus said. "And I was in love. I was just in love," she continued. Others seem to love it, too—she describes it as a "hit" at the bar—but the chef knows that while the taste is enticing, the part that has made the beverage so irresistible is the arresting aesthetic. "And that was the beginning."

Although kudzu lemonade is a seasonal staple at home.made, Maumus's kitchen has since whipped up other kudzu-based treats, with the chef trying her hand at kudzu jelly and a kudzu crème brulée. Whenever she thinks up a new way to create something delicious with the plant, she gives it a whirl. The taste and texture aren't the only allure—the southern legends surrounding kudzu add an enthralling dash of mystique. "I've always just seen it as something magical," she said. Now fifty-two, the mind behind the drink that's captured the hearts of many Athens residents has an attachment to cooking from wild things that stretches back to her youth. Born and raised in New Orleans, Maumus grew up in a household where preparing meals wasn't just something you did to survive, but a cherished practice that was integral to everyday life. "I would go fishing with my father, come home, filet the fish in the backyard, and then see it turned into this beautiful meal," she remembered. It wasn't something her dad did for a living, but something he drew immeasurable joy from.

A contagious love for the culinary craft wasn't something she inherited just from her family patriarch. An adolescent Maumus learned about canning and making chutneys, pickles, and jellies, among other delicacies from her grandmother, who had a garden teeming with sources of food. Being around these regular practices where gathering and preparing food was both a wellspring of joy and a staple of a healthy routine made a distinct impression on the chef. "It was the difference between people who are making a meal to feed the family to get it on the table versus the people who are enjoying that process," she observed. Maumus would go on to work in restaurants before pursuing other fields of study in college. She would come to find she couldn't stray far from the gourmand habitat she'd grown up in: "I missed the environment. I missed the culture. I missed playing with food." It wouldn't be long before an innate call to cook sent her down a path back to the restaurant industry. She would ultimately move to Athens and work under the tutelage of now-renowned restaurateur and *Top Chef* judge Hugh Acheson. While working in one of Acheson's kitchens, Maumus simultaneously launched her own catering business from her house—which is where the moniker "home.made" would come from. Eventually, she and a few friends rented a tiny kitchen space in a ramshackle part of town to expand their offering, with the very first reviews dubbing the street the "Boulevard of Broken Dreams." Maumus eventually stepped back from her work at Acheson's restaurant to go all in with her own endeavor. The chef saw the culmination of a dream when home.made first opened its brick-and-mortar location in 2006, but the opening collided with an unforeseen life event, something that would change her forever.

The morning after the day home.made first welcomed in the outside world, Maumus got a call from her sister, who had devastating news—their father had passed away the night before. He had died the very night that her restaurant opened. According to Maumus, her dad had spent more than twenty years of his life in a wheelchair, lacking mobility and the ability to "feel anything from his chin down." Because of that, he wasn't able to travel to Georgia to attend home.made's grand opening—something that Maumus vividly remembers he was upset over having to miss. Reeling from the sudden loss, the chef found a poignant comfort in the realization that her father was able to attend, after all. "I believe in the universe and I believe in magic and I believe in all kinds of beautiful things," she said. "And I feel like my dad was there. He wanted to be at the opening, and he was there." Limitless love and grief fueled that moment and will buoy the chef through all of life's challenges to come. "Thankfully just being gifted the perspective of 'My dad's here with me', that really helped me get through that," she affirmed.

That was over a decade ago. In the years since, people have traveled far and wide to experience home.made, beckoned by classically decadent southern food with a modern spin, served in a rustic, cozy ambiance. Maumus even launched "Sidecar," an adjacent jazz bar in the same building as the eatery. And then came 2020. COVID-19 descended on the world, sending officials and scientists scrambling for answers. As everything screeched to a stop, local businesses shuttering left and right, home.made grappled with the financial blow that would derail Maumus's entire industry. Restaurant and food-service sales fell by $240 billion in 2020 alone. Pivoting to catering, Maumus's company stayed afloat—despite the impossible odds stacked against it. Maumus likes to say they are as resilient as a legion of mature kudzu. "I guess I'm kind of like kudzu in that way," Maumus concluded. "I just kept creeping."

Her initial entanglement with the "vine that ate the South" predates her work as a professional chef and restaurateur. After college, Maumus didn't head straight to Athens to set up shop. At first, she found herself in a pocket of western Georgia's countryside, trying her hand at adulthood. It was there, in the "teensy little town" of Roopville, Georgia—which had a population of just 231 in 2020, according to US Census data—where she first came across the incomparable plant. What Roopville lacked in people it made up for in kudzu. "I would drive past this patch on the side of the road that was just trees, completely overcome by kudzu. It really was just so beautiful to me," Maumus said. She doesn't remember if it was flowering, drawing her to its seasonal array of blossoms. What hypnotized her was the sheer abundance of the plant—and how it sparked a childlike imagination, one that conjured up visions of mythical creatures and beguiling beings. "These huge forms that were covered in this ivy-like grass that I would look at and I would imagine 'Oh, there's a T-Rex,' and 'There's this,' and I would just imagine these storybook scenes," she noted. "It was just like cloud-gazing."

An almost immediate captivation with an unknown plant, leaves cascading like a waterfall across the sites it claimed as its own, is what Maumus most clearly remembers. "I thought it was completely beautiful," she said. It was years later when she would learn more about the vine that captured her fancy. "I didn't know it was invasive. I didn't know any of that stuff. I certainly didn't know it was edible," she confessed. She couldn't help but be enthralled by the visceral response it evoked, thanks to the sheer growth it appeared capable of. "I thought Coca-Cola should do a commercial with something coming out of the kudzu," she added. "It looked really magical." This sensation stayed with her, becoming far more than a passing notion about a plant casually observed on the side of the road. A thought that would never really disappear, rooted, in many ways, like a deluge of kudzu embracing an abandoned plot of land, deep in the rural countryside. "There were times that I thought, 'Well, there's so much that I wish someone could figure out a good use for it," she said. The chef says this stuck with her, especially in light of the constant shortages of food affecting people across the world. "I wish this thing could be helpful, because there's so much," she remembers thinking. "What could we do with it?"

That is exactly the question she has funneled into her own craft. It's no surprise that the chef and mastermind behind home.made's menu is obsessively interested in constantly innovating ingredients and techniques. "I was naturally on this path toward unusual ingredients," she said, "and just always asking, 'Is that edible?'" It was how she'd eventually find her way back to the plant that once caught her eye. After seeing it at local farms, her curiosity was, once again, piqued—leading the chef to want to cook with it. "Kudzu, good lord, almost every part of that thing is edible," she proclaimed. At first, she tried to experiment with the tap root, but couldn't get any local farmers to give her a large enough root to be processed into a powder. Knocked down but not defeated, Maumus would turn to the leaves—toying with different ways of preparing them. It wasn't long before kudzu lemonade was born. There was no set plan that triggered it, but rather a result of a sudden inspiration—a rekindling of a spark that had inadvertently lit once upon a Georgia road.

If you ask Maumus, where and when you gather what you use in the kitchen matters. She is quick to warn other chefs interested in cooking with wild edibles that it's important to be careful about where they are sourcing the vine—it makes sense that collecting kudzu off of roadways or plots of land right by industrial facilities will expose it to toxic exhaust—and what time of year they're collecting it, as well as what types of leaves to look for. In her case, she prefers to use smaller leaves from younger stands. But she's also not out there pulling kudzu herself; every fall, professional foragers collect copious amounts of it for home.made. Still, every now and again, she'll pick some from her very own collection. "I have it growing in my backyard," she said. "My daughter loves it. The kudzu lemonade got her hooked."

Raised in Louisiana, Maumus didn't grow up exposed to any charged stereotypes about kudzu—that understanding would come later, when she moved to the Peach State. The way she sees it, the older fables and poems depicting the weed as a monstrous green being are rooted in misconceptions. "It is a beautiful green monster, though," she stressed. "I've just always said it's like the clouds. Why would you get rid of the clouds?" She knows kudzu is a threat to native ecosystems but argues that all it takes is awareness and effort to rein it in. The property she lives on cradles a substantive sheet of the herculean plant, which Maumus trims and watches out for overgrowth so it doesn't overcome her adjoining vegetable garden. She's not bothered by the work that goes into that, though. For Maumus, the aesthetic appeal and utility of the vine more than make up for the maintenance. As far as the chef is concerned, kudzu firmly has her "Georgia heart." "Considering how much we have of it and how easily it grows, if you want to talk about a renewable resource, we have one that we're not using," she noted. Of course, kudzu is not the only invasive with untapped resources in the South. Maumus believes the utility of other weeds like magnolia is overlooked by most—in everything from culinary to textile applications. "These things are thriving for a reason.

And I think oftentimes, they're there for us," she said. "And if we could figure out why it's there for us and use it for us, then we could learn to really love it." This could be an offshoot of a wider modern disconnect with nature and all it has to offer. "Our relationship with the natural world, it tells us things. It informs us. We just don't listen," she continued. "If we could find a way to listen instead of pushing it away and shushing it, and instead saying, 'Well wait, why are you so abundant here? What's so great about here?'"

There's a chance that we've been looking at kudzu all wrong. Maybe the very nature of an invasive vine is a message to humans from nature itself—a startling reminder that embracing how our environments and ecosystems are adapting to a warming world is the key to surviving it. "Maybe kudzu is trying to say, 'Here I am. Here I am to remind you that I'm a gift. Figure me out, and use me,'" said Maumus.

"Maybe it's *The Giving Tree*[2] and we just keep cutting it down."

---

2. Shel Silverstein's *The Giving Tree* (2014) is a popular children's book about selflessness and sacrificial, unconditional love.

6

COLLECTING AND CONNECTING

One idle afternoon in Gulliver's Cove, a quaint fishing village surrounded by acres of dense woodlands in southwest Nova Scotia, conservation biologist Joe Roman found himself immersed in an engrossing hunt for European green crabs, one of the most fabled invasive species. It was this on-the-ground exploration that led him to a chance meeting—one that, like most seemingly innocuous encounters, would change everything.

While walking along a rocky shore, Roman came across a weathered, scruffy man carrying a five-gallon bucket with miniature holes cut into the bottom. For a few moments, Roman watched the stranger collecting something indiscernible from a rock-strewn beach, quietly observing as what appeared to be pint-sized snails sporadically fell through the vessel, ricocheting back onto the sand below. Something about the way he searched for his mystery catch didn't strike Roman as the typical manner of a biologist. As his curiosity piqued, Roman walked over to where the newcomer stood and asked what he was doing. It was there, at the place where the coastline of Gulliver's Cove meets the Bay of Fundy, that he would learn something that would trigger a series of events that transformed his life. It turned out the stranger was harvesting European periwinkles—an invasive species in North America—to sell in markets in Boston and New York. The encounter revealed to Roman the rising popularity of such catches in the cooking world; doubling as principal elements of Italian and Chinese cuisine, the green crabs and periwinkles were, at the time, high in demand among restaurants in larger metropolitan areas. "He didn't know that they were invasive, or care. He was just trying to make a buck from collecting the species during low tide," Roman recalled. There, on the shores of Nova Scotia, a man collecting invasive crabs to make a living sparked the flicker of a wild but overwhelmingly logical idea: What if there was some way to encourage the incorporation of invasive species and plants on the menus of trendy eateries across the US? Could this help mitigate the spread of invasives everywhere?

This all transpired in 1999. Roman, then a thirty-something, was pursuing his PhD in conservation biology with a specialization in marine ecology. His thesis focused on the European green crab, which was introduced to the United States in the 1800s, likely first brought over from Europe by hitching a ride on sailing ships. Initially spotted off the coasts of New York and southeastern Massachusetts, the diminutive shore crab's destination would eventually be the scenic shoreline habitats lining the Cape Cod peninsula—one of the very first US destinations settled by colonists nearly two hundred years prior. So it came as a shock to marine biologists when the European green crab was mysteriously spotted in 1989, far from the northeastern shoreline where it was left. That year, those very same scuttling sand-dwellers were identified more than three thousand miles away, in San Francisco. A 1998 study by Andrew N. Cohen and James T. Carlton published in *Science* found that the San Francisco Bay and Delta area's accelerating invasive species rate, in part attributed to commercial shipping, made it one of the most invaded estuaries in the world. The UC Riverside Center for Invasive Species Research cites over 175 "exotic" species established in the tidal area, and 75 "exotic" species established in the Sacramento–San Joaquin Delta, a watershed known as the "hub of California's water supply." Over the next decade, the invasive marine invertebrate spread north from the Bay Area—popping up in Oregon, Washington, and finally, British Columbia. Ranked number 18 out of the top 100 "World's Worst Invasive Alien Species" by the Global Invasive Species Database—whereas kudzu is number 77 on that list—the European shore crab has built up a distinct reputation as a species that disrupts existing ecosystems within salt marshes and pocket estuaries, as well as impacting the commercial shellfish industry. Devouring other invertebrates, the crab consumes large quantities of native mussels, oysters, clams, and even other shore crabs.

Just ten years after the European green crab had been spotted in British Columbia, Roman found himself scouring shorelines to collect the overabundant crustaceans for his research. From Provincetown up to Nova Scotia and Prince

Edward Island, his expedition would lead him to drive and stop every fifty kilometers to flip rocks and look for crabs along northern Atlantic coastlines. While investigating the coastal predators' invasion record, he stumbled upon something surprising. Until then, the marine biology world had been under the impression that the invasive species had spread north from Massachusetts. What Roman had discovered was a different origin story—he had traced the crabs to another invasion stemming from northern Europe in the 1980s, which introduced it to Nova Scotia. That under-the-radar invasion was the driving reason behind the omnivores migrating all the way up to Canadian coastlines, rewiring the biodiversity of an area that had never before seen its kind.

Alongside the millions of European green crabs that had descended upon the Nova Scotia area, the common periwinkle also littered the intertidal zones of the Gulf of Maine. Not to be confused with the purplish-green flowering vine—which is, incidentally, also an invasive in the US—the common periwinkle, *Littorina littorea*, is a snail species that scientists believe arrived in the region as early as the 1800s. A 2009 study published in *Proceedings of the National Academy of Sciences* led by University of Maine marine biologist Susan Brawley found they were likely transported by ballast rocks on board ships traveling to New England from Scotland and England. These deceptively prosaic-looking snails are attracted to rock-strewn beaches, where they feed on algae adorning rocks and seaweed. It is widely believed that they significantly alter the diversity and abundance of the intertidal ecosystems they invade. In the Gulf of Maine, an overabundance of common periwinkles has been found to displace native snails, as the non-native snail can eat algae found on the coastline faster than its native competitors, forcing other snails to search for food elsewhere.

Invasive problems identified, Roman set out to collect more than just the sea creatures he'd been originally searching for. Periwinkles joined the soft-shell green crabs in his bucket, before later landing together in a frying pan. The result was an explosion of flavor. And just like that, he was hooked. "I've spent my life trying to reduce the impact of humans on wildlife. Many fish are overfished. Many species are overharvested," he explained. "And really, at that point, I'm like, 'Well, here's a species that I would love more people to collect.'" Armed with an expanding roster of invasive species, Roman realized that finding European green crabs was no longer his sole objective. He was encouraged by the economic interest in collecting invasives he'd witnessed from the nameless fellow he'd met on a stretch of Nova Scotia's beaches. That was when he set out to try a version of his own, one with a significantly more sustainable objective. His approach expanded beyond green crabs and periwinkles, opening up to include all invasives—even kudzu. Having lived in Florida for a period—he got his master's degree at the University of Florida—the biologist was quite familiar with the perennial vine, which abundantly adorns freeways and the margins of forested land across the state. "After that, after I had started thinking of this idea, I called up a bunch of chefs around the country," he said. This was in 2004. Roughly two decades later, his crusade to collect unwanted species for human consumption is well-established nationwide, thanks largely in part to *Eat the Invaders,* Roman's digital publication where you can find hundreds of recipes for cooking with invasive plants and animals. Over the last ten years, as the movement of "invasivorism," or eating invasive species in order to reduce their populations, has picked up traction, the work of *Eat the Invaders* has been profiled in outlets like the *New York Times,* the *Guardian, NPR,* the *New Yorker,* and the *Washington Post.*

"I'm a biologist, not a chef," affirmed Roman, who teaches about invasives at the University of Vermont. "Not only am I interested in conservation, but I'm interested in good food and thinking about new ways of teaching people about the ecological history of their area."

Invasives à la carte is just one avant-garde way he unified the two worlds, one offering a thrilling glimpse into a vast expanse of knowledge about what precisely is on a plate, and the other an enticing foray into dazzling new flavors, textures, and tastes. Kudzu became a gateway into edible invasive enlightenment. In the midst of launching *Eat the Invaders,* Roman came across José Gutierrez, a Memphis-based chef who was working with the invasive vine in his kitchen. The conservationist reached out to the chef, procured Gutierrez's recipe for kudzu sorbet, immortalized it on *Eat the Invaders,* and just like that, everyday readers suddenly had what they needed to whip up ice cream à la kudzu at home in the comforting glow of their familiar kitchen spaces. If you compare the taste of the perennial vine to the abundance of other invasive species harvested on *Eat the Invaders,* it still finds a way to remain distinct. Although many other invasives rely on garlic, onions or olive oil to make the flavor burst, kudzu is a different story. In more ways than one, the vine stands apart from its peers. "Kudzu is actually lending the flavor. Its flower is a flavor enhancer," explained Roman. "On the other hand, if you're looking at the starches, then it's going to be what you add to that starch. So in other words, if you add the roots, that is going to be what you add to it."

Eat the Invaders often approaches controlling invasives, like kudzu, from a native perspective, imitating what works and what doesn't. "Of course, eating a couple of kudzu blossoms, well, it's fun, and it tastes good. And I hope it would expand awareness. But it's certainly not going to reduce the impact of kudzu in the South," Roman said. "There are also recipes for the root. And there is increasing acknowledgment that we can use this for furniture, or artwork, or architecture." In pockets of East Asia, kudzu is a native plant, and the applications of the vine across these industries is immense. "Really, the approach for kudzu would be thinking about how it's used in Japan, thinking about how it's used in Asia, and then you can adapt it for the US," he proposed. This was precisely the idea behind ringing up as many American-based chefs as possible in his ecological pursuit. Ever since, the *Eat the Invaders* approach has been broadly adopted for several invasive and non-native flora and fauna across the US: lionfish, invasive carp, and even dandelions.

Belonging to a large genus of flowering plants, common and red-seeded dandelions are wildflowers native to Europe, thought to be purposefully cultivated by settlers to the US in the 1600s, with some lore suggesting that they were brought aboard the *Mayflower.* Ever since, dandelions have spread across the country as a weed. Indigenous communities and European colonists both incorporated the flower into everything from medicine to food, practices that are still common today. Unlike kudzu, ecologists say the blooms pose little threat.[1] Roman's own family once sought out the easily harvestable dandelion. "My great-grandmother from southern Italy used to collect it every spring," he recalled. According to Roman, one way of getting people interested is to look at the culinary background where the species is native and then adapt these recipes to North America. He noted this could very well be the case for southerners who incorporate common East Asian uses of kudzu into American-style dishes. "You're bringing it to the South," he said.

When he first proposed the idea of equipping chefs with the recipes and materials they need to incorporate invasives into their menus, Roman was met with utter silence. "The response was crickets," he declared. In the decade

---

1. They are, however, classified as invasive in Alaska, where the common dandelion reigns supreme on the state's targeted plant species watchlist as of 2022. "Invasive & Non-Native Species," National Park Service, February 24, 2022, https://www.nps.gov/subjects/invasive/ak.htm.

since *Eat the Invaders* launched, that momentum has dramatically changed. In part, he credits the foraging movement for the budding attention. Five years after he first started calling chefs up, challenging them to think creatively about invasives, all while asking for recipes, Roman began to notice a shift in the responses he was receiving from culinary connoisseurs. Suddenly, he was getting callbacks. At last, the people he had been tirelessly reaching out to were curious to learn more, drawn in by the novelty of the idea and the promising ecological upshot. If America's foraging movement continues to grow, Roman suspects there will be an increased interest in eating invasives as a means to help protect and preserve the environment. "You've got the foragers, and you've got the restaurant tours," he said, "and that can result in a change, which we've seen."

Many modern wanderers have made their way across forests, farmlands, and even urban landscapes scavenging wild plants, fungi, berries, and nuts as food sources. This personal collection of plants has been around as a source of survival for as long as humans have. We know it as foraging, and we don't need to live in the countryside to do it. A 2017 study in *Urban Forestry and Urban Greening* found that foraging in urban cities is becoming more popular because of the human need to connect with nature. It has further amplified since, as food insecurity, or a lack of access to nutritionally adequate food, has increasingly become a reality for millions of people across the world. Grappling with the COVID-19 pandemic, which led to lost jobs, supply chain disruptions, and escalating prices, further exacerbated by the war in Ukraine and severe climate events, the 2022 United Nations report on food security and nutrition found that about 29% of the global population is at least moderately food insecure.

Across the US, rising rates of food insecurity have been a significant stressor that disproportionately impacts communities of color, people with disabilities, and the unhoused. Christianna Silva reported for *NPR* in 2020 that 19.1% of Black households and 15.6% of Latino households experienced food insecurity the year before, which was nearly double the 7.9% of white households who were food insecure in 2019. Meanwhile, climate change has produced unique challenges for Indigenous communities from Alaska to New Mexico, who face escalating volatility in the subsistence food systems they rely on.

It's a damning reality that has led some to look for unorthodox solutions—like urban foraging. Eloquent with a lilting, South African accent, when Marie Viljoen talks, her words meld together like spoken poetry, enkindling scenes bound to send you to another place. Intentional and expressive, the urban forager is skilled at making the experiences she's lived come alive for others. Viljoen is based in New York City, a predominantly industrialized landscape. She's the author of *66 Square Feet: A Delicious Life* and *Forage, Harvest, Feast: A Wild-Inspired Cuisine,* as well as a forager, blogger, published author, and contributor to several digital outlets. Her wild-food recipes have made her prolific on Instagram, where over 38,000 people follow how she turns the edible plants and fungi she collects from city parks, private lands, and wild areas around her home in the Big Apple into delicacies. "The real appeal of foraging is a sort of self-sufficiency," Viljoen said. "It's giving you a sense of control over something . . . you're going out hunting for something, and if you find it, there's a real sense of achievement, and 'I've done this.'"

It's a takeaway that makes a lot of sense, as a loss of control in this day and age seems to be more prevalent and persistent—especially as people living in nations like the US grapple with dangerously heightened political polarization. Amid the lasting toll of the COVID-19 pandemic, devastating school shootings, deepening social divides, and worsening climate change, it can feel as if the world as we know it is crumbling, our sense of faith in prevailing justice forever lost. Like a house built with cards, our foundation is far from indestructible. "People are feeling that they're not in control of a lot," Viljoen added.

Viljoen's backstory is easy to relate to. It's an age-old fable, one where we stray too far from our path before life sends us back to our roots. Her mother Maureen was a professional gardener, meaning the younger Viljoen was raised in a household where she became well-versed with the language of the land early on. "I was a very, very young child when my mother essentially just gave me seeds and sent me into the garden," she recalled. "Probably just to get me out from underfoot." Coming of age in Cape Town, South Africa, she recalls that her childhood was filled with planting herbs, flowers, and vegetables, like the radish seeds she had been given to play with in a serene garden oasis, carefully cultivated by her horticulturist parent. Endless days were spent familiarizing herself with the living things emerging in her backyard botanical wonderland and memorizing the Latin names of plants. A teenage Viljoen first stumbled upon designing gardens, not long after also falling passionately into cooking. "It's hard to separate plants and food in terms of which was a bigger passion," she said. "From a very young age, they've been intertwined."

Viljoen left that path professionally—for a beat, she studied opera. "It should probably have been botanical," she reflected bemusedly. "But it took life experiences to bring me back to gardening, and garden design, and foraging," she noted. Reconnecting with her curiosity—for plants, for cooking, and for conservation—the creator found her way back to the elements of her life that had brought her infinite joy. And there she remains. "I'm so connected to wild growing things because it's something I do every single day," she said. "And I sort of need to be around plants."

Listening to the forager talk reverently about the nexus within creating our own food from edible plants, sometimes gathered from our own backyards, is akin to unwrapping a gift you didn't necessarily ask for but instantly adore. Viljoen's fervor for making meals from resources drawn from the environment, and even more specifically, plants otherwise overlooked, is contagious; the way she describes what she does is invigorating to even the most nature-averse among us. "My way of creating is to cook, and being able to draw on unusual plants, flavors, spices, herbs that are not conventionally available is very exciting," she explained. "It's almost like a painter having access to colors that other painters don't have. It's a cook drawing on unusual ingredients that most other cooks don't have access to, simply because they're not aware of them." Tapping into a rare knack for storytelling, one that ignites interest in the secrets hidden in the habitats around us, Viljoen leads foraging classes and walks across New York. In her spare time, she carries on one of her adolescent hobbies—designing gardens. Although she finds the full scope of her work difficult to describe, if you ask her, she will tell you she is professionally and personally obsessed with plants. "It's very hard to describe what I do in a simple sentence, so I'm a little self-conscious about labeling myself in so many terms, but whatever I do is related to plants, essentially," she said. "The foraging, the garden design, and the writing, all tend to be about plants."

Especially those that can serve as sources of food. Launching a garden-centric blog sometime between 2008 and 2009, Viljoen was already drawn to the promise of edible plants—intrigued by the way they've remained steady fixtures, crucial human customs throughout time, a mosaic depicting the ways our ancestors embraced the vastness of the earth. As a cook and creative person, she recalls being fascinated by the connection between everyday cooking and everyday plants, beyond a garden setting. She remembers discovering that so many plants that are considered weeds in the US are in fact common ingredients in other parts of the world. "It became this very curious conversation of culture versus food versus 'What does this plant mean here?' And that was a slow evolution," she observed. Viljoen makes it a habit to seek out invasive plants when she forages, and she incorporates those into her recipes. The pull for her is the conservation benefit, although she knows eating invasives isn't likely to eradicate them. "You can't only mitigate by eating a plant. It's really hard to eat them into submission," she noted. "But increasingly we are [becoming] aware that the herbicides we are using, even the ones that are now so-called 'better' than the previous ones, are having a quantifiable effect on the environment, and it's worth pursuing alternative ways of control." Over the past decade, a growing body of research has demonstrated the disastrous effects of some commonly used pesticides on human and ecosystem health. In 2020, researchers at McGill University found that the mass-market herbicide Roundup could trigger biodiversity loss in freshwater ecosystems. And a 2022 study in *Environmental Health* led by a team at George Washington University found that one in three Americans has been exposed to weed killers like 2,4-D, which have been linked to detrimental impacts on human health, ranging from hormonal imbalance to cancer. That same year, the US Environmental Protection Agency announced plans to begin assessing the potential effects of new pesticides on plants and animals protected by the Endangered Species Act.[2] According to Viljoen, what we already know about the repercussions of herbicides should be enough to pursue more unconventional methods of invasive management—including nature-based strategies like foraging. "Essentially foraging is a form of mechanical control of invasive plants," she said.

She's come across kudzu in the New York City area just once. Blooming and boisterous, the vine stole her attention, leading her to collect some of its blossoms to experiment with. "I've only used it in terms of its flowers because that was the time of year I stumbled upon the vine in the Bronx. And it was like the 'vine that took over.' I could totally see why that's kudzu's name," Viljoen

2. The act does not specify explicit protections for fungi, which has long been met with criticism. Only two species of fungi have ever been listed under the federal legislation. See Naveed Davoodian, "A Long Way to Go: Protecting and Conserving Endangered Fungi," New York Botanical Garden, October 15, 2015, https://www.nybg.org/blogs/science-talk/2015/10/a-long-way-to-go-protecting-and-conserving-endangered-fungi/.

recalled. The blossoms she found became a filling for homemade summer rolls, a Vietnamese dish traditionally made from rice paper wrappers partnered with fresh vegetables. "The flowers were amazing," she said. "They were very delicious, pea-like, sweet and super versatile."

People's expressions change when she brings up plants like kudzu, largely in the context of how they are invasive to the US and native to East Asia. "There's a slightly frozen look," she noted. "Because nobody wants to hear that a person from Asia, for example, is considered unwanted or undesirable in the United States. So, we are constantly having this multi-level conversation." She doesn't think she's ever had a class or taken people on a foraging walk that didn't feature the subject in some way. "I hear that frequently . . . that [people] equate native foods with Native people, and invasive plants with derogatory language about people," Viljoen said. "We get into this use-of-the-philosophical talk about it because plants and people do behave very differently. But, in this country, where the conversation about race is very fraught, I have found it being applied to plants." Getting firsthand accounts of how a plant that grows in the New York Botanical Garden is commonly eaten in Japan, where any one of her walking-tour participants may have been raised as children, is something she considers immensely valuable. "It does also just make me, as a person, more sensitive to the immigrant experience," she continued. "Which is very different from my immigrant experience."

Born and raised in Constantine, South Africa, Viljoen immigrated to the US later in life, eventually calling Brooklyn her home. She acknowledges that her experience assimilating to life in America hasn't been rife with the same rhetoric or prejudice faced by immigrants hailing from predominantly non-white cultures. "As an outsider, it's always interesting having a slightly different perspective, and simply because I sound a little bit different and am perceived as an outsider. At the same time, I'm a white woman. So I sort of fit into the white culture," Viljoen said. This carries a different weight today, as immigration and America's exclusionary border policies remain hot-button political topics across the nation. "I'm not what some of my neighbors would consider an immigrant, because I look like them, which is a whole other story," she said. While she thinks it's an important conversation, the forager sees the need to draw a line of distinction between humans and invasive plants. "It really forces you to reevaluate plants and people, and the whole conversation about who comes from where, and why biodiversity is a good thing in a natural situation," she added. And yet, an invasive plant like kudzu does directly damage the ecosystems it invades through its capacity to outcompete native vegetation, which ushers in a whole host of ecological problems. Typical questions that emerge in her line of work include: "Why do I want to preserve the native plants in a certain forest? Why am I against barberry, which is an invasive plant, growing in the forests in upstate New York, for example? Why is that bad?" From there, those attending Viljoen's foraging walks usually plunge into a conservation conversation.

She'll hear the most about these societal nuances from the younger crowd who join her classes and tours, everyday people who are eager to examine why ecologists use the language that they use to describe harmful plants with origins from elsewhere. "Unless I have someone who has a Native American background on a walk and then they say, 'Well, colonists were invasive.' And I'm like, 'Yeah, that's a very good point,'" she agreed. According to the urban forager, you can't escape the cultural associations embodied by plants. You can't have one without the other. They're as intertwined as kudzu vines, spiraled into the soil they feed and feed from. "It's a constant," Viljoen said. "You're constantly having the human conversation while you're having the plant conversation."

Tama Matsuoka Wong agrees. At sixty-three, the New Jersey native forages for a living. Thirteen years ago, she left behind a career as a lawyer to found "Meadows and More"—her company that supplies foraged wild edibles to wholesalers and restaurants across New York City. "I'm coming about it from a forager's point of view. It's just, like, my wacky, wacky way," she said. "Why are all these names uglified? Why is it Japanese honeysuckle, or Japanese knotweed?" Wong finds it odd that the classifications of invasive plants wouldn't just give invasive plants similar names to what they're called in the countries they are native to. "A lot of invasives are Asian, especially Japanese," said Wong. She's a second-generation American from a multicultural household; her dad is Japanese, and her mother is from Hawai'i with Chinese ancestry. Her mom is an "ultimate New Age Hawaiian cook," while her dad is a "scientist with this Japanese love of nature."

Born in Japan, Wong's dad is familiar with plants in the US that are native to his home country, like kudzu. She says her dad even taught her that kudzu—

or kuzu, as it's referred to in Japanese—is named after a prince.[3] In the early 2000s, Wong moved back to New Jersey after spending several years with her family living in Hong Kong. It was around then that she tried her hand at planting and growing her own garden. "We did not have kudzu in New Jersey," said Wong. That has since changed—which could have something to do with the rapidly warming climate. Nowadays, she says it has been spotted in the Garden State. As of 2019, thirty-six populations of kudzu had been identified in New Jersey. Amid these reports of the unwanted vine creeping in, Wong began hearing from a handful of people in New York who were looking to collect it. One such request came from a group hoping to source kudzu to replicate a conventional design strategy in India: using palm leaves to make tableware. By then, the state's Invasive Species Strike Team was mapping emerging invasives in order to eradicate them before they became a problem. Wong got the team to share some of the kudzu they had cut down with the people looking to create with it. "They were cutting down this stuff anyway, right? So they were cut down, they brought these garbage bags, and then I delivered it to the people in New York," she recalled.

Unlike Chef Maumus, Wong is not a "kudzu in the kitchen" proponent. In fact, she questions why any chef would use the vine in their culinary creations. After she expressed interest in trying the notable weed, friends in Atlanta sent her part of a kudzu leaf. "And it doesn't taste good. At all. It's really thick and hairy," she observed. She hasn't yet tried the blossom, however—although she hopes to one day. "I'm very interested in this, because it's never flowered here," she said, referencing how it's used in Chinese herbal drinks—like five-flower tea[4] or in the aptly named kudzu juice[5]—and for alternative medicinal appli-

---

3. The origin of the plant name in Japanese is said to have connections to royalty; *kuzu* in Japanese also means *katsura*. In 1645, the Katsuras were members of Japan's imperial family, and the Katsura Imperial Villa in Kyoto, Japan, was built for Prince Toshihito in the early Edo Period. See Albert J. Koop and Hogitaro Inada, *Japanese Names and How to Read Them: A Manual for Art-Collectors and Students* (London: Eastern Press, 1923), https://archive.org/details/japanesename showookoopuoft.

4. Five-flower tea (五花茶) is a traditional Chinese herbal tea that cleanses, soothing inflammation and sore throats. It is made of *Pueraria lobata* (kudzu), honeysuckle, chrysanthemum, silk cotton, and *Plumeria rubra* (frangipani). See "Traditional Herbal Teas in Southern China," n.d., http://www.shen-nong.com/eng/lifestyles/chinese_tea_traditional.html.

5. A 2020 article in *Time Out Hong Kong* describes kudzu juice (葛菜水) as a tea that soothes inflammation and aches, as well as cases of dry mouth. See Ann Chiu, "Chinese Herbal Tea in Hong

cations. "It's only [used] with specific other things that are supposed to have a medicine or benefit. It doesn't mean it tastes good," added Wong. From the perspective of someone who works to supply restaurants with wild edibles, taste is the number one thing. Although mitigation or control of invasive species growth is a benefit of foraging, what's equally important to Wong is identifying and selecting plants that are appetizing—invasive or not. "You can have tons of things that are edible, but they taste like cardboard," she said. "If you want people to get excited to get rid of all the kudzu, people aren't going to be like, 'Wow, they said you can eat it, but it was [like] cardboard.'"

That's where the professional forager hesitates to recommend the use of invasives like kudzu in culinary applications. "[It's] combined with these other things according to Chinese history, right? But you can't say it tastes good for modern restaurants," she said. Another factor is abundance and ease of collection—Wong says she works with invasives "if there's a lot of it, and I can get it." In other words, if it's not readily accessible where she is, it's not a priority. She has yet to come across any kudzu where she forages. "I'm not trying to walk five miles and find someone morel[6] in the woods," she said. Where Wong lives is akin to a forager's oasis, a vast expanse of land in Flemington, New Jersey, brimming with at least 225 wild plant species, the majority of which are native. In the time she has lived there, she has mitigated all the invasive plants that have threatened native populations. "If you harvest it for ten years, it's kind of gone," she declared. Wong has worked with conservation groups across the state trying to get rid of a myriad of invasive plants, including Japanese honeysuckle. "Although 'get rid of' is also relative, as you never really get rid of it. But if you keep doing it every year, you can make a good way of managing it," said Wong.

Lately, mugwort and brassica rapa are two of many wild edibles that Wong finds herself drawn to. These changes as the seasons do, her interests charged by what blooms around her. What's crucial to Wong is her approach to foraging—and the message her work sends. "I am approaching this from an environmental perspective," she explained. Adding to that, Wong has an intrinsic connection to the plants she spends her days growing, collecting,

---

Kong 101," *Time Out Hong Kong,* January 2020, https://www.timeout.com/hong-kong/health-and
-beauty/chinese-herbal-tea-101.
  6. A type of wild mushroom.

and experimenting with. She doesn't understand objections to anthropomorphism—or attributing humanlike characteristics to nonhuman life forms. The way she sees it, how else would we be able to define and describe these life forms, the multitude of plants and animals that we interact with on a daily basis? "We only have words as humans. We're limited by language," she said. "You can't say personality because that would be crossing the line in terms of attributing human characteristics. [But] I think they have an intention."

Plant neurobiology, the study of the behavioral patterns and characteristics of plants, is still a relatively new field, with the first meeting of scientists who research it taking place in 2005. A 2021 *BBC Science Focus Magazine* article by Efraín Rivera-Serrano dives into the results of a 2020 paper published in *Scientific Reports,* which analyzed the bending movements of French bean plants in a type of behavioral analysis of their growing patterns. When comparing the growth of bean plant shoots that were potted with and without vertical cane, the study authors found that the plants potted with a cane would grow "along more predictable paths." While Rivera-Serrano reported that the research doesn't prove that the plants acted with conscious intent, it does demonstrate that the plants in question were doing more than merely responding to the environment and stimuli around them.

The deeper we look at that idea, the more unsettling it can be. There's something to that discomfort, equally mystifying as it is expected, which a few researchers are probing despite a bevy of criticism waged by peers across the world. A professor of forest ecology at the University of British Columbia, Suzanne Simard is one of the most widely known names trailblazing this work. The first of her landmark ecological discoveries was published in 1997 in the scientific journal *Nature* when she revealed that when trees die, they communicate information and send carbon, sugars, water, and nutrients to other trees as well as other plant species. Such a thing is possible through elaborate underground root systems that connect trees and soil fungi to one another by what is known as "mycorrhizal networks."

In Simard's book, *Finding the Mother Tree: Discovering the Wisdom of the Forest,* the ecologist illustrates what her research has unearthed: Plants appear to have agency, and the largest, oldest trees are "mothering their children" through the ways they send and share information and resources to younger, smaller trees. This suggests their ability to make decisions, learn, and remember. *Finding the Mother Tree* focuses on amassing evidence pointing to how "Mother Trees," or the largest, most ancient trees in a forest, do this. And it's not something she identified only taking place among trees; it requires the active participation of soil fungi species, too. A 2021 *Scientific American* interview with Simard by Richard Schiffman breaks this idea down perfectly: "You found that birches give sugars to fir trees in the summer through the mycorrhizal networks and that firs return the favor by sending food to birches in the spring and fall, when the birches lack leaves," Schiffman said.

When she first published that 1997 paper, much of the response was heavily charged and highly negative. More than a quarter of a century later, voices of dissent linger but have since been minimized and muted, as Simard's research has proven "beyond doubt"[7] that these complex root and fungal systems not only exist but are used by trees in the way humans cooperate in social networks. Brought to vivid life in *Finding the Mother Tree,* these systems in the soils of the forests and fields and plots of land around us should be understood as parallels of human ecosystems, Simard argues; trees are social creatures just like we are, and this should be interpreted as a sign of intelligence. A common criticism is that trees and other plants don't have brains— which Simard's ongoing investigations may someday end up refuting. Looking at the belowground structure of a plot of trees, she sees an aesthetic similarity to that of a human brain and has said in interviews that they have structures that are very similar to it.

Major questions concerning the implications of intentionally eradicating plants—such as something so universally condemned in the US as kudzu— have begun to emerge in the wake of scientific discoveries like these. For now, these probes lay motionless, buried in the nutrients flora and fauna thrive in, stymied by stereotypes and unspoken rules in the world of science. Just another set of borders erected in the name of humanity's relentless pursuit of knowledge—lines etched into a set system, some never intending for them to be crossed.

Of all the words to describe plant activity, Wong says the one she likes best is "choice." "Plants choose to do what they do. I don't have a better word for

7. This was said by Krista Tippett on a 2021 podcast episode of *On Being with Krista Tippett:* "Suzanne Simard—Forests Are Wired for Wisdom," *The On Being Project,* September 9, 2021, https://onbeing.org/programs/suzanne-simard-forests-are-wired-for-wisdom/.

it," she affirmed. It's not that she thinks plants choose the same way humans do. The way she sees it, plants are not sitting around, wearing "thinking caps." Critics of the idea of plant consciousness chalk actions that appear to be an intelligent choice to nothing more than genetic encoding and outcomes of natural selection. Despite that, the wave of scientists in opposition are unable to combat the work of ecologists like Simard, who, step by step, are bringing forward a profound idea that challenges our understanding of plant ecology: Forests do not behave as many organisms, but merely a single one.

"Obviously, plants, they choose to grow someplace or not grow someplace, right? You can try and have some plant grow somewhere," Wong said, "and it just doesn't want to be there."

7

A HOUSE IS BUT
A HOME

For some of us, it takes a lifetime to discover what we're meant to do. Others figure it out early on—led to their calling as if by magic. At age seven, already obsessed with making things, Kyle Schumann learned how to use an old bandsaw in his family's basement, a lesson provided by his equally crafty father. "My mom was not happy when she learned about it," Schumann laughed. Much to her chagrin, from that moment on, he was resolutely hooked; never still, he could always be found tinkering with all the tools he could get his hands on. "I wanted to be an inventor," he said, "which has come full circle, actually." As he grew into the sustainable architect and designer he is today, Schumann fell into his intended career path swiftly and naturally, almost like one falls in love. "The overarching theme for me personally has always been just making and building, and how we can use different tools and technologies to transform the materials around us to be useful and productive."

Now an assistant professor of architecture at the University of Virginia, Schumann is one half of the duo that runs After Architecture, a design initiative Schumann founded with fellow architect Katie MacDonald over a decade ago. Together, they work on sustainable architecture projects, guided by a vision of one day building a supply chain of construction materials from invasive species. Experimenting with repurposing invasives to build a more ecologically friendly world also stems from an early childhood influence for Schumann. Growing up on the grassy pastures of Vermont, he shared his backyard playground with dozens of alpacas; his parents operated an alpaca farm. As a teen, he learned how to collect fibers from an alpaca's fleece and how resources like an animal's fur can be ethically incorporated into practical uses. "I've always been around or aware of different sorts of natural material streams. . .and how those are used by us, by people," Schumann said.

In many ways, Katie MacDonald, the other half of After Architecture, has a similar backstory. As a kid, MacDonald was always busy creating something. "I'd make personalized versions of board games and personalized nutcrackers and all these kind of bizarre things," she recalled. "You just have all that creative energy and want to make something and share those things with other people." She met Schumann while they were undergrads at Cornell University. Both studying architecture, they teamed up for a few student competitions, which led them to learn how fervently they each felt about experimenting with an untapped realm of the building design space—natural materials. Ever since, the two have worked on what they call "ecologically positive" projects together, trying to innovate natural material solutions to the world's largest crisis—climate change. And they're doing what they set out to do, too. One build at a time.

By taking on design projects that aren't emissions-intensive, the After Architecture studio works to lessen the construction industry's material role in contributing to climate change. For MacDonald, it's especially personal. "That mission for me became more urgent in 2018, when my parents lost their home, as well as neighbors, in a series of events that resulted from the Thomas Fire in California," she said. In December 2017, the Thomas Fire burned about 281,900 acres and destroyed 1,063 structures, damaging 280 more, before triggering a series of flash floods and mudslides the following January that wrecked 400 houses and killed twenty-three people. "I've seen my home state of California transform with these fires of radical scale. I mean, it just became enormous and the impact has been huge," she added. "And it also just became really personal. So the climate crisis hit home for me at that time, and just made the work that much more urgent."

Everyone everywhere is on a similar trajectory. For most of us, it won't be long before the magnitude of the impacts of climate change will morph from stories read on a page to experiences lived out in real time. In fits and waves, it's already happening—rapidly rising temperatures and intensifying extreme weather events are becoming increasingly impossible to ignore for all of the world's population. This is clear across the Global South, as the emerging era of coinciding disaster events is driving people to flee their homes, from the Philippines to Somalia to Guatemala. But it's not just disproportionately affecting residents of developing countries—the consequences of climate change are being felt by people occupying every nation on the planet. It's evident in Europe, amidst shattered temperature records and summer heat waves that stole the lives of thousands. It's omnipresent among arid landscapes found across the United Arab Emirates, where climate change is a threat multiplier, fueling some of the world's worst water crises. It's unmissable across Oceania, as severe bushfires incinerate forestland and sea level rise compounding with coastal erosion threatens the future of low-lying coastal nations. No region will be left untouched, no community unscathed. And at the rate humanity is combusting fossil fuels, no person will be able to escape it. Millions are already being forced to move in "the Global Climate Migration," as a growing wave of climate refugees are left to fend for themselves in a search for safety.

Warming temperatures are deepening racial and social divides, too. People of color are suffering the most and the worst consequences of these extreme climatic changes. In the US, where essential elements for survival—like healthcare, housing, and food—are not a right, but a luxury, communities of color are left to bear that unjust burden. A 2022 study in *Current Environmental Health Reports* compiled evidence of this, with researchers finding that Black, Latino, Indigenous, Pacific Islander, and Asian communities in the US are at a "higher risk of climate-related health impacts than Whites." Their peer-reviewed work effectively makes the case that climate change is nothing short of an environmental injustice, likely to exacerbate existing disparities in everything from mental health to diseases—and, in the worst of cases, even premature death.

Whether you're in California, where warming temperatures are fueling more intense and frequent wildfires, or Florida, where current sea-level rise estimates predict parts of low-lying coastal Miami will be submerged by the close of the century, or anywhere in between, the direct landscape effects are obvious. The indirect fallout is more implicit, but it's there, lurking in the background, like the monstrous vine of nursery rhymes that ceaselessly "haunts" the South. Nearly all industries and economies are threatened by climbing temperatures; agriculture, health, and infrastructure are at the top of that list. A 2019 study in *Nature Climate Change* analyzed twenty-two sectors of the US, finding that under a high emissions scenario, the nation stands to lose an estimated $520 billion every year by the end of the century.

Meanwhile, leading fossil fuel companies are the predominant contributors to more than one-third of the world's greenhouse gas emissions in the modern era, further exacerbated by carbon-emitting corporations in industries such as building and construction. And the built environment makes up nearly 40% of total energy-related carbon emissions worldwide. "We're interested in how the materials we work with engage the efforts to mitigate climate change," explained MacDonald. It's something other designers are considering, as alleviating the effects of the climate crisis has become a larger part of the construction conversation. A year ago, MacDonald kicked off a research initiative investigating how various grasses can be assembled into sustainable materials. Teaming up with a few fellow UVA professors—one specializing in environmental science, another in composites engineering, and a third in structural engineering—in a project affectionately nicknamed "Mass Grass," MacDonald is exploring the radical question, "How do we make grass into a structural building material?"

While MacDonald has been busy looking into the untapped potential of pastures, Schumann has had another focus: the bright, beguiling world of "green" tech. For the last few years, he has concentrated on advancing sustainable technology that could create utility out of otherwise unused invasive species. Joined by fellow UVA faculty members specializing in environmental science and mechanical engineering, it was in the fall of 2021 that Schumann launched "Variable Sawmilling," a research project that innovates invasive trees. They're figuring out how to create the type of technology that would robotize a sawmill—something that could be tapped to process invasive trees that otherwise can't be removed, transforming them into usable material in a way that historically is often difficult or impossible to do. A lot of invasive species might be "too bendy" or have different material properties that require specialized methods of cutting to make them usable, according to Schumann. One that his team is looking at is the *Ailanthus altissima*, or "Tree of Heaven." A deciduous tree that originated in China, not only does the Tree of Heaven spread very rapidly, leaving disrupted native species and ecosystems in its wake, but it is also a common host for another infamous invasive pest and substantial ecological concern; the spotted lanternfly.

An abandoned barn overwhelmed by kudzu in 1946.

(Carol M. Highsmith Archive, Library of Congress, Prints and Photographs Division.)

A swath of kudzu enshrouding trees and a telephone pole in Tate County, Mississippi.

(Ben May Charitable Trust Collection of Mississippi Photographs in the Carol M. Highsmith Archive, Library of Congress, Prints and Photographs Division.)

This haunting shot renders a gas station overtaken by kudzu. Such a place doesn't
actually exist; in fact, it's nothing more than a product of Casey Lance Brown's
creativity, the result of manipulated images of several gas stations
fused with large-format imagery of a hillside consumed
by the vine close to where the artist lives.

(Casey Lance Brown, from Kudzilla series, 2020–. Courtesy the artist)

Featuring images like this one, where a stand of trees is consumed by sheets of kudzu,
Casey Lance Brown's visualization series on the "scourge of the South"
relays a pointed message about the multitude of ways one
invasive plant has been miscast in popular culture.

(Casey Lance Brown, from Kudzilla series, 2020–. Courtesy the artist)

From Sears to GNC, logos of major companies litter this arresting visual of a roadside teeming with the "vine that ate the South."

(Casey Lance Brown, from Kudzilla series, 2020–. Courtesy the artist)

The word "DECAY" illuminates a mass of kudzu overtaking an old wooden building, framed by a thicket of trees blending into an almost imperceptible skyline.

(Casey Lance Brown, from Kudzilla series, 2020–. Courtesy the artist)

There's something eerily familiar about this depiction of a plot of land consumed by
an endless sea of kudzu. By casting a neon pink layer over the visual,
artist Casey Lance Brown conjures up a strange new world
that still somewhat resembles our own.

(Casey Lance Brown, from Kudzilla series, 2020–. Courtesy the artist)

Abandoned objects peek out from underneath a veil of kudzu.

(Casey Lance Brown, from Kudzilla series, 2020–. Courtesy the artist)

Plant ecologist Heather Coiner works in a field swaddled with kudzu in Leamington, Ontario, one of Canada's southernmost municipalities.

(Image courtesy of Heather Coiner.)

Every piece crafted by Alabama artist Beth Phillips incorporates kudzu vine. Lamp bases are upcycled from discarded or damaged lamps that have been disassembled and repurposed.

(Image courtesy of Beth Phillips.)

A ten-by-ten-foot installation created by Katie MacDonald and Kyle Schumann was featured in the 2020 exhibition *Homegrown* at the Knoxville Museum of Art. The architects used kudzu, bamboo and various tree species, as well as forestry waste to build the structure's biocomposite panels.

(Image courtesy of Katie MacDonald and Kyle Schumann.)

Four walls form an exterior room, roughly the size of a standard domestic space, according to *Homegrown* designers Katie MacDonald and Kyle Schumann.

(Image courtesy of Katie MacDonald and Kyle Schumann.)

First appearing in the US in 2014, the fabled lanternfly feeds on the sap of more than seventy plant species, inflicting serious damage to native trees and crops and leaving behind a sugary waste that promotes the growth of sooty mold as well as attracting other insects to it. The USDA National Invasive Species Information Center classifies the lanternfly as an "economic threat" to US industries like "viticulture, fruit trees, ornamentals and timber." It is native to China, India, and Vietnam. In July 2021, *The Conversation* published a piece by University of Tennessee professor Frank Hale that chronicled the path of the spotted lanternfly invasion. Hale wrote that the lanternfly "had spread to about half of Pennsylvania, large areas of New Jersey, parts of New York state, Maryland, Delaware, and Virginia." Last updated on December 1, 2022, the New York State Integrated Pest Management map documenting the lanternfly's spread shows it has also been seen in Massachusetts, Connecticut, eastern Ohio, and even as far west as Indiana and north as Vermont.

"Tens of millions of dollars of agricultural damage are going to be caused by this insect as it flows through the South, and the Tree of Heaven is a primary host tree for it," noted Schumann. "If we look at how to make this tree usable as lumber, then that is providing some economic incentives that you could work with the Department of Transportation and other places to incentivize removing the species." By removing the Tree of Heaven from the US ecosystems it invades, the rate of spread of the spotted lanternfly should also decline in the places lacking an abundance of host trees for the pest to feed on. Schumann wants people to understand how this work is relative to urban timber movements—a cultural renaissance where felled trees are harvested by local woodworking facilities as a method to lessen a city's carbon footprint. As communities become more resourceful, the architect believes the demand for a sawmill that can cut down invasive trees could grow, especially as it becomes more commonplace across the country to reuse trees that might come down in a cityscape during storms or disaster events instead of simply tossing them into landfills.[1] If successful, his team hopes to have some impact on lessening the rate of spread of the spotted lanternfly. But Schumann's goals stretch beyond that milestone; he wants to challenge expectations of traditional architecture, redefining what is considered standard within the building industry.

1. According to Sam Sherrill in his 2017 book *Harvesting Urban Timber: A Complete Guide* (Echo Point Books & Media), between three and four billion board feet of lumber from urban trees in the US ends up in landfills every year.

"As a society, we have ideas of what we expect architecture to be, and this is why we're developing these research projects to have some built prototypes, which are really important for us to show people," he asserted. "Things that might be radical, or things that you might not normally expect. Like a house made of grass."

Or kudzu. According to the designers, kudzu is a hard fibrous material—meaning it's both tough and flexible, two elements that make it advantageous to build with. Although it can't replace traditional building materials like copper or aluminum on its own, when combined within a composite panel, alongside other elements like unwanted timber, the weed makes for a worthy substitute. They also say that using the plant in architecture is a newer method in the industry—as of 2023, they don't know of anyone else trying to build with it.

It's a little bizarre, given that the utility of the vine can be traced back more than two thousand years in Eastern culture, where records show it was used as a source of fibers for construction materials, among other purposes. In 1665, fiber derived from the vines and stems was even incorporated into the production of cloth and paper. But the use of kudzu in Western construction hasn't been widely recorded. In the US, MacDonald and Schumann are the first to have created and displayed an architectural assembly made predominantly from parts of the weed. Their use of the "scourge of the South" isn't accidental. "In specifically targeting kudzu as one of the species to use, we're trying to question preconceptions about what architecture is or what architectural materials could be," explained Schumann. He talks almost reverently about kudzu's resilience, and how its complex American footprint could be translated to the application of the vine as building material. "There's a cultural power in trying to tackle kudzu," he continued. "The impacts you can have, in terms of how people perceive it right now as a nuisance or a barrier in the landscape . . . that could be transitioned more toward something potentially useful, or a possible resource."

Their pursuit of creating with kudzu began several years ago. When MacDonald and Schumann were architecture fellows at the University of Tennessee in 2020, the formidable entity was everywhere, embalming the fringes of the land where they lived with layers of glossy leaves. "It's hard to avoid it, and you see it blanketing just about everything. It becomes a real presence in the landscape," MacDonald noted. She remembers how impressed they were by the sheer abundance of it, creeping up and down power lines and mailboxes as if it were alive. "It really transforms certain parts of the landscape into things that almost look like a Dr. Seuss illustration. There was something alien, or surreal, about kudzu that struck us immediately," she added. As designers, they saw this overabundance, and the local dialogue around it, as a possibility. There seemed to be an opportunity space where they might be able to incentivize something that's good for the environment, like remediation, by making it a useful act of building material, according to MacDonald.

For Schumann, the dual nature of the vine is what makes it so compelling. "There's this perception, on the one hand, that kudzu is a horrible thing in the native ecosystem and taking over," he expressed. "On the other hand, there's a really beautiful quality about it. Something so foreign from the landscape that you wouldn't see otherwise, that maybe can be appreciated for that, even though you're understanding the other ramifications of what it means for it to have taken over an area." But swapping out traditional construction materials for invasives as a possible emissions-friendly solution to design projects doesn't come without challenges. "Using kudzu panels as structural systems for a building, there's a lot more that you have to grapple with in terms of proving that the system works, it can be safely engineered and is up to standing, what fire rating they need to ascend and whatnot," Schumann admitted. "But if we're talking about using them as some finishes within a project, then there's some more leeway there." Although he and MacDonald are still trying to figure that application piece out, they confess that they are endlessly intrigued by the material-sourcing puzzle of something as abundantly accessible as kudzu. "You're taking care of two things at once, right? You're removing the material, which has to be done to restore that native ecosystem, but you're also then putting that material toward some useful purpose," he said. "It's not like it gets trucked away and you have to pay to dispose of it."

These days, the two are busy working on proving that they can feasibly build something with the iconic weed. To a degree, they already have. Constructed in 2020, Homegrown was a 10-foot by 10-foot domestic room made out of kudzu, bamboo, and forestry waste, which spent some time as a temporary installation at the Knoxville Museum of Art. It's just one example of how the designers are working to reframe their industry into something climate-friendly. They dream of doing more. At one point in the future, they'll be kicking off construction on a 17-foot-tall, 13-foot-square pavilion made entirely of panels built from kudzu and bamboo plants. It'll be bigger, bolder, and hopefully bring even more attention to their ecologically positive cause. They're also actively looking to file a patent for the technology they invented that will help conceptualize this elusive kudzu-based installation. But it's not just about using kudzu or other invasive species—it's about all local materials that are available to us, in our larger, exceedingly precious biosphere. "If we can convince people that you can use kudzu, you can overcome all of that stigma. There's also a whole spectrum of things in between kudzu and traditional pine lumber that could also be productively architectural," said Schumann—before noting a vested interest in the "larger argument" of how natural materials are used.

Together, they're creating a supply chain of building materials for the future, as they fight to do their part in preserving it. It's a cause that architect Joseph Valenti is also working toward. In 2018, Valenti made headlines across Hawaiʻi for his postgraduate project—constructing an affordable housing unit made entirely of albizia, an invasive tree.

Born and raised in Southern California, Valenti has lived in Hawaiʻi for the last fifteen years. After moving to the island chain in 2006, he went on to pursue a doctoral degree in architecture, where coursework would eventually send him to China, a country infamous for its unprecedented urban growth.[2] Valenti credits his experience living within some of China's most renowned cityscapes as what guided him to the idea of centering invasive species in design concepts. Only recently has the nation started pursuing sustainable "green" approaches to its built environment, joining a trickle of other wealthier countries that are facing social pressures to take part in a global trend of eco-friendly construction, in response to the ballooning ramifications of climate change. Seeing the "scale and destructiveness of the built environment over there" and "how much the landscape has transformed so quickly" convinced the architect-in-training to rethink what elements can be used in design processes.

"I felt, when I came back to Hawaiʻi, a new sense of importance of *where*

2. Out of the world's one hundred largest cities, about one-fifth are in China. See David Satterthwaite, "The World's 100 Largest Cities from 1800 to 2020, and Beyond," International Institute for Environment and Development, January 16, 2020, https://www.iied.org/worlds-100-largest-cities-1800-2020-beyond.

our materials come from, especially being on an isolated island," said Valenti. Hawai'i has also experienced an "urban explosion" of sorts—in 2014, 92% of the state's population lived in an urban area. Since the 1800s, steel and concrete have been found among the most commonly used construction materials on the archipelago. It was a reliance he became hyper-aware of in his own industry. "Being surrounded by that, and then being isolated, led me to looking deeper into what resources are here locally," he recalled. A fixation with sustainable building using natural resources culminated in a doctoral project, when he decided to build a tiny housing unit made entirely of a locally invasive tree. The architect was trying to produce a viable solution to two distinct issues—an overabundance of invasive species and a lack of affordable housing. Hawai'i is one of the most expensive states to live in, facing the most inflated housing costs nationwide. A reliance on imports for building materials is one of the factors driving up residential fees. This tied into the essence of Valenti's ambition to build using locally sourced materials. "It was a combination of, 'Let's see if we can look for local resources to build with,' and also applying that to one of the needs in the community here," he said.

The genesis of this two-pronged venture was another research project hosted in 2015 by the school where Valenti was completing his postgraduate degree: an albizia removal at the University of Hawai'i at Mānoa's arboretum, which doubles as a public botanical garden tucked into the back of Mānoa Valley. In the thick of nearly two hundred acres of tropical rainforest on the island of O'ahu, Valenti witnessed the removal of a stand of albizia trees. Watching in awe and perplexity as those mighty sources of lumber were chopped down and discarded in a ravine led the budding architect to consider an uncharted alternative to what he viewed as excessive waste—creating building timber from felled, invasive trees. The spark of an idea kindled into a full-blown inferno, leading Valenti to develop a prototype on the campus of the University of Hawai'i at Mānoa School of Architecture. It was a more than $80,000 process that would be funded by the US Forest Service, the Hawai'i Housing Finance and Development Corporation, and a grant from the school itself. What he didn't expect was the hurdles to come—things that would heavily complicate what started as a seemingly straightforward task. Using invasive trees in this first build was a major feat, made all the more significant by elements out of his control, from a lack of available, usable resources to a lack of time.

Valenti now admits that if he had known what he would face to make that vision a reality, he probably wouldn't have gone after it. "I got a lot of push-back. Just coming up with the idea and proposing it," he recalled. The first hurdle to clear was getting everyone around him supportive of building with albizia. A haunted legacy in Hawai'i is to blame for that—the tree outcompetes native vegetation by blocking water sources and has the reputed ability to grow up to fifteen feet every year. Much like kudzu, it was introduced by a person pursuing unexploited answers to other problems. Botanist Joseph Rock established albizia populations in 1917 to help reforest Hawaiian lands that had been stripped by cattle farming. Valenti thinks the initial resistance, from peers to professors, stemmed from an instinctive aversion to something famed for disrupting native ecosystems and entire forests across the state. But he wasn't looking to plant more of it—just harvest what the archipelago already had in abundance. Part of that is a vision, a vestige of hope, that if the work becomes popular enough, an invasive species with origins in Indonesia and Papua New Guinea might one day be eradicated from Hawai'i.

Once he was able to convince all naysayers the build wouldn't contribute to the local albizia predicament, but would actually help minimize it, the real work began. It was 2016. Collecting materials for the venture was chaotic and complicated, both time- and labor-intensive, sending Valenti and his collaborators to a slew of removal projects across the island of O'ahu, including some at the site of a US army barracks and off a highway in Mililani. The blueprint mapped out the prototype to be built entirely from albizia wood, modeled in the vein of a tiny house, landing at roughly four hundred square feet.

Then tragedy struck. Without warning, Valenti's mom fell into a coma. Pressing pause on the work, he went back to Southern California to be with family. There he would remain for six months as her condition deteriorated. The following year, without waking up from her coma, Lisa Valenti passed away. Awash in grief, the promising architect would head back to Hawai'i. There, his project lay frozen in time, progress halted by an unforeseen loss. But the deadline to complete it hadn't changed. It was imminent. Racing against the clock, Valenti and his team completed most of the build in less than forty-eight hours, leaving parts of it unfinished, just in time for spring. Etched into the final full-scale structure was *Lika,* the Hawaiian translation of *Lisa.* A home named in honor of the person who would always represent his.

It hasn't been replicated since, although not for lack of trying. "I've wanted to evolve this to something that can be applied," he noted. The media and academic attention that emerged around the construction process would lead to the dawn of the Albizia Project in 2017, an organization headed by Valenti that has since developed into a team of four, who have created everything from surfboards to rolling cabinets—all using their namesake tree. But albizia is not a code-compliant building material in Hawai'i. Without that official stamp of approval, the Albizia Project can't use the lumber source beyond demonstration or prototype capacity—much like the model Valenti completed, which allowed them to bypass building material code restrictions. "We haven't built anything in a structural sense, but if we were to start building structural albizia homes or affordable housing with albizia, we'd have to go through a pretty rigorous testing process," he explained. Although the Hawai'i Invasive Species Council encourages residents to get rid of the albizia tree, the verification process of the invasive plant as a possible building material will take years of vetting before it may be a feasible formula to eradication and involved in widespread housing designs.

Funding is also blocking the path forward. Sourcing and working with albizia trees as a main building material is a task that comes with a hefty price tag, a bill that Valenti and company haven't been able to foot on their own. Lately, their greatest roadblock has been finding contractors willing to take on the cost of figuring out how to make albizia an approved building material. "We are just not in a position to fund that. And we haven't gotten anyone wanting to jump into that with us yet," he said. When it comes down to it, the architect knows this particular unit design concept isn't likely to be widely applied because of that. "What we did was preliminary," he affirmed. "It was just to prove that it can be done."

This could change down the line. Although this particular trend of sustainable design is rather new, there's international support—it's a movement that is emerging, flourishing, and picking up traction as more of the world works to lessen the planetary impact of one carbon-polluting industry. "If you break down one of our biggest carbon footprints in society, it's our building industry," he observed. "If you look at where most materials are coming from, in most buildings, they're definitely not benefiting climate change." In his eyes, wood and invasives are a great substitute for steel and concrete. Combining the two—like, say the invasive albizia tree—is ideal. "I'm an advocate for wood as a building material, in general, just because it is a renewable resource," he added. "Taking it from an invasive is scoring a few steps beyond that, because we need to remove this invasive that's costing the state in numerous ways to try to manage it." Valenti hopes to see more of his peers centering sustainability in their designs—including those that incorporate invasive plants like the albizia. At the very least, he wants to see a concerted effort to reconfigure an industry so egregiously reliant on burning coal, natural gas and oil.

If architects and designers were able to transition away from steel and concrete, and even use lumber that is sourced locally, Hawai'i—and the broader industry in general—would experience significant reductions in overall carbon emissions, according to Valenti. By that logic, using albizia wood, which he notes sequesters carbon, would allow for builds to contribute less to climate change. Meanwhile, repurposing an invasive tree would also help restore native habitats. "That's our big vision," explained Valenti. "If we can start to remove albizia, find a use for it, and then help restore the forest that albizia is overtaking, the native species or beneficial species to Hawai'i's ecosystems, then we're tackling climate change in multiple ways," he continued. Common criticisms circling the idea of basing something economically profitable on the abundance of a detrimental species—with the end goal of eradication—aren't lost on Valenti. Some of the questions he has heard before include: What happens if building with albizia is so wildly successful, the tree is wiped out of Hawai'i's forests once and for all? Isn't there a dangerous chance of commercializing something that could end up with people actively planting and spreading the invasive species, to meet an industrial supply and demand? Thankfully, the answer isn't as stark as, "We'll cross that bridge when we get to it." The architect wants to assure any skeptics that these are questions he and his team have given a lot of thought to. "If we were to, say, start manufacturing a furniture line that's going to use a pretty steady supply of albizia, I think it would be decades out until we actually felt we were running low on this resource," he said. "We're specifically just taking unwanted albizia that, in most cases, is already being removed."

But that's beside the point. If they're successful with their initiative, which he notes would include working alongside Native Hawaiians and community organizations who share their objective of restoring native ecosystems, he sees the end result as a net positive. "In twenty, thirty years, if we do start to diminish the supply of albizia, then we've already planted the seed for other species to be harvested as a new resource," he noted. In the not-so-distant future, that's their aim, according to Valenti—a sustainable supply chain of building materials that use invasive species as a means to mitigate and eliminate the forces that are threatening and altering native ecosystems. Part of the reason his team has been working with the forestry industry in Hawai'i has been to highlight and demonstrate the potential of invasive species in builds beyond their installations. They've designed a site for a future build they call an "accessory dwelling unit," which he describes as a small-scale housing unit constructed entirely from eucalyptus and pine harvested from a local plantation. If you ask the architect, he'll tell you all about his hopes that the movement they're heralding is catching on, noting how it's becoming "more and more mainstream in the industry to be mindful and conscious of how we're building and what we're building with."

Of course, Valenti's pleased to see that slow-burn intrigue climb as awareness of climate change intensifies—not just among his peers, or within architectural circles, but in every industry and aspect of human life. For Valenti, each successful build carries with it a reminder of the ways we can work together to rehabilitate the harmful impacts of human industry.

"It's a beautiful solution, because people are starting to see there's a story to it," he said. "But there's also something special about seeing this material in, whether it's your home or the space that you're building, instead of something that was brought in from, most likely, a not very sustainable forest. So the meaning of it is important."

8

ROAD TO RECOVERY

Snaking through the outskirts of Dobson, North Carolina, belts of one awe-inspiring weed blanket a plot of land the size of a football field. The plant's bristly, yellow-green stems and sporadically heart-shaped leaves mean it is often mistaken for a colony of greenbrier or poison ivy. But it is neither of those. This is kudzu, an invasive species that is known to smother its victims by way of murderous limbs. Contrary to popular belief, kudzu doesn't choke out other vegetation until it ceases to exist, like female octopi who strangle their partners while mating. Its vines wrap around other plants, sheets of leaves blocking out any sunbeams and effectively cutting off their food source. For any being dependent on photosynthesis to survive, no sun means no shot at survival. By that logic, kudzu is a death sentence.

Every August, Tamera Wilson toils across an overgrown space along that rural North Carolina road. Day after day, she braves the oppressive heat and persistent threat of bee stings to fill bag after bag with kudzu blossoms, which she and her husband Darryl stockpile and use to make kudzu-based jams, sauces, and dressings. They join a growing list of people across the Southeast working to repurpose invasive species like kudzu. The hidden forces driving them vary—some are interested in financial gain, others hope to encourage embracing natural resources, and many are passionate about eradicating plants that disrupt biodiversity through eating, healing, or building with them.

The Wilsons are the founders of Carolina Kudzu Crazy, a business Darryl and Tamera launched in Thurmond, North Carolina, after they went from using kudzu to feed their farm animals to producing a line of kudzu-based products. The year was 2011, and they were seeking a new direction. "We were looking for something, praying about different things that we could do and get involved in," said Darryl. Rewind ever so slightly, to a time when the two, while cultivating a humble plot in the Blue Ridge Mountains of North Carolina, embarked upon some at-home research to find more cost-effective ways to feed their amalgamation of rabbits and pigs. It was then, at an otherwise ordinary moment, when the Wilsons first stumbled across mentions of an alleged source of protein embedded within kudzu's trifoliate leaves. Curious to test that for themselves, they collected some of the weed, which happened to engulf the edge of their property.

"I went out one August, pulled out a kudzu vine, filled up my bag, and harvested it for our rabbit," Wilson recalled. After looking into kudzu further, he came across an old cookbook with a kudzu jelly recipe in it, which led him and his wife to whip it up and give it a try. At first, they were not fans—the taste was sharp, fundamentally bitter, all shades of wrong. But the couple just couldn't ignore an instinctual feeling that they were onto something. That hunch is why they kept tinkering with recipes, dovetailing a myriad of flavors and ingredients, until they, at last, perfected the art of cooking with kudzu, producing everything from sweet to spicy glazes and jams based around the vine's blossoms. Then, the world got in the way. Before they could develop their recipes even further, the vine's growing season disappeared with the first signs of winter frost, putting their plant-powered plan on pause. The following August, once it began to bloom again, the Wilsons jumped into action, setting out to harvest as much of the weed as they feasibly could. A few dozen jars of kudzu jelly recipe later, the couple had innovated the "perfect" blend of kudzu blossoms. "We gave it to people at church, and people started to buy it!" Wilson exclaimed. Just like that, they had inadvertently created a demand for their kudzu jam in their close-knit community. An idea started to take root: What if they started to sell these kudzu-based blossoms to the people they knew? That year they started slowly, like most do with any sparkling new venture, promoting it only among family and friends. They didn't have to wait long, though—it was clear from the community's response that what they had made was nothing short of a vine-flavored hit.

Then 2013 and 2014 rolled around, bringing an unexpected insect invasion that halted any line of thought on expanding the business. The dreaded, formidable kudzu bug popped up in the area where they collected the vine, gobbling up the blossoms and leaving the Wilsons without the core ingredient of their new product. But Darryl and Tamera Wilson aren't quitters, so they held out hope that they would eventually be able to get back on track. And in 2015, once the last of the bugs were eradicated, the Wilsons found themselves eagerly returning to the grueling business world of kudzu. Reinvigorated by their insect-free kudzu patch, the two created even more kudzu jelly and began appearing at small-scale community festivals with their newfangled culinary creation at hand. Before they knew it, they were making hundreds of jars a season. As demand swelled, the two felt the pressure to come up with a name for their booming local business. In 2017, while the husband-and-wife duo was devoting all of their time to developing even more kudzu-based products, "Carolina Kudzu Crazy" was born.

It was only when they decided to make it official that the Wilsons hit another obstacle—the state. After contacting the relevant agencies about being licensed, they sent their recipes and products for inspection and testing. But the state sent it all back and "refused to test it," according to Darryl. Dejected, but not giving up, he reached out to the US Food and Drug Administration (FDA) for advice. Because of kudzu's role as an invasive species, he was told to "give up" the pursuit. Meanwhile, the Wilsons were also advised that they wouldn't be successful in getting their business licensed.

Undeterred and hell-bent on succeeding, the couple connected with an official at North Carolina's Department of Agriculture and the USDA to plead their case. There they found an unlikely ally: Steve Troxler, the head of the North Carolina Department of Agriculture, who told them that if they could help him find someone in the country who had credibly researched kudzu, he would push it through and get the product tested. They searched and searched, eventually unearthing a handful of people fitting the mark, but when they tried to reach out, they learned some had passed away. Others were not responsive. But their commitment to the thorny dilemma would be fruitful in the end. At long last, the Wilsons connected with someone who had personally conducted peer-reviewed research on the safety of kudzu-blossom consumption. With evidence in hand, they were able to get the state to reconsider their application. Officials tested the recipes, came out and inspected the kitchen, and finally said they could move forward. It had been a year and a half since they had first reached out about getting licensed, but the battle wasn't over.

According to the Wilsons, officials at the FDA had told them the process to get licensed would take a decade—and require capital the family just didn't have. "[The FDA] said it will take about ten years and $10,000," Darryl noted. That was the final tipping point, the proverbial hammer that shattered a not-so-distant dream. With little hope left, the Wilsons prepared to give up and walk away from their aim to rewrite the culinary narrative with the vine that consumed the South. After all that time hand-collecting and innovating with kudzu, the inherent force propelling them forward started to vanish, like a bloom crowning a flowering weed in the fall. That is, until Troxler, their ally at the North Carolina Department of Agriculture, suddenly got back in touch with some welcome news. "He said that $110 was all we had to spend, and they were going to send it on," Wilson said. "So we sent it in and we got everything straight. For $110." By the time 2019 rolled around, Carolina Kudzu Crazy was a legitimate business in the state of North Carolina, with the Wilsons managing to do it in one year instead of ten.

Ever since, they've diversified the products they offer, moving beyond jams to developing kudzu-based sauces for barbecues and stir-fry recipes. Balsamic vinegar dressing à la kudzu blossoms is one of their latest concotions. Nowadays, Carolina Kudzu Crazy is still a team of two—made up of Darryl and Tamera—but they are no longer limited to operating during the plant's growing season. Taking a page from traditional tribal subsistence stores, they stockpile and freeze kudzu blossoms, which allows them to cook them, freeze the juice, and produce their swiftly burgeoning line of commodities year-round. It takes a cup of kudzu blossoms, each usually the size of a dime, to make a cup of their jelly. "You can imagine how much picking that is," Darryl said. "We harvest pounds and pounds and pounds of the flowers every year."

In 2022, the Wilsons sold roughly five thousand kudzu-based products. Given kudzu's outsized degree of cultural infamy across the South, Wilson admits they have consistently faced community resistance in their embrace of the despised vine. Initially, they only distributed to people they knew directly, and they didn't face any pushback. In fact, everyone they first introduced Carolina Kudzu Crazy to was nothing but positive about it. Things only got dicey once they expanded into broader customer territory. It was a "mixed bag" when they officially launched, according to Wilson. "Once we went out and sold it in more of a public setting, we found a lot of people that were so against kudzu that they didn't even want to look at it," he said.

But with time comes wisdom—as well as more refined sales tactics. What they've since uncovered is that the vine's problematic reputation, and the visceral reaction of many southerners to it, is also a talking point, a connective tissue entwining people together. The stigma that follows kudzu-centered products often means that at local and regional events, where the Wilsons get most of their sales, strangers will come up to their station and eagerly offer complaints, sharing anecdote after anecdote about their lived experiences battling the demonized vine. The founders of Carolina Kudzu Crazy embrace that form of inbound marketing; they'll tell you that by engaging with those skeptics, they can be converted. "They come up and they walk by and they say all kinds of stuff," Darryl said, citing the countless times he's been told how terrible the weed is, or how people remember being forced to go out and pull the roots out when they were teenagers. In moments like those, Wilson jumps at the opportunity to share how he and his family learned to champion the vine, giving others the chance to see it as something more.

Originally from the West Coast, the Wilsons moved to North Carolina in the early 2000s. A landowner himself, Darryl understands the resistance to the idea of reframing an invasive species in a positive light, especially the risks associated with commercializing it. To that point, he stresses that he and Tamera get the entirety of their kudzu blossoms from one plot of land—their neighbor's two acres of kudzu-overrun fields. Despite harvesting it all in one place, Wilson claims that they still haven't used more than one percent of the kudzu growing in that solitary space. "We don't even make a dent," he asserted, with a shade of pride and unspoken reverence.

Alongside community events and festivals, at least five stores across North Carolina currently sell jars and bottles of sauces steeped in a wily, mystical plant courtesy of Carolina Kudzu Crazy. Word of mouth is their biggest advocate. Leaning into that, they've even started selling merchandise: classic t-shirts and face masks (manufactured in light of the COVID-19 pandemic) printed featuring a homage to kudzu. Growing demand is nearly outpacing them. "We realized this year that we could not harvest enough to meet all the demands that we're looking at," explained Wilson. The husband-and-wife duo is now looking to hire someone to help with sales. They have enterprising, longer-term dreams that they're slowly working toward. "Our goal is to hopefully one day make it into Cracker Barrel," he laughed. In part, their ambitions rely on the cultural recognition of the wild ingredient at the heart of their company. "I want my name to be synonymous or right there beside something like Smucker's," he declared.

It's not just a case of capitalistic opportunity for the family, either, as they'll tell you they don't simply harvest, package, and market products starring kudzu. Over time, the inimitable vine has become a staple in their everyday lives. "We do things with kudzu for ourselves and our kids and grandkids. We fry them up and make chips, make a substitute for spinach because my grandkids hate spinach. And make cookies," he said. They also occasionally make wine with it. But it's not just for consumption; they have crafted napkin holders, kitchen utensils, and jewelry using parts of the weed. One year, the Wilsons even—quite proudly—created a Christmas wreath using kudzu. "We use the vines to make baskets and things like that. I make natural-looking birdhouses that hang on a tree, like a nest with a vine," he said. Everything the fifty-four-year-old does is for his grandchildren. They are the driving reason why he works as hard as he does, constantly pivoting and scaling up the services and spaces dominated by Carolina Kudzu Crazy. By teaching his grandkids to tap into the wonder of the precious, wild world around them, he hopes to leave behind a legacy where "misjudged" sources of utility are embraced.

"I'm not that old yet. I still have a few years left," he noted, musing on what's to come. "I want to do my part and turn it around and show people that they can benefit from this abundance that we're blessed with. Even though it looks like a curse."

The storied perennial might be considered a plague in Western civilization, but it's not viewed with the same sense of derision elsewhere. In China and Japan, two of the main places the vine originated, kudzu has been embraced for at least a millennium as a core healing ingredient used in treating ailments such as digestive issues, itchiness, dizziness, nausea, headaches, and high temperatures. Records in Chinese tell of kudzu roots being dried and diced for medicinal purposes as early as 1578; others suggest treatments using the vine go all the way back to the age of the Western Han Dynasty, between 206 BC and AD 8, as recorded in *Shen Nong Ben Cao Jing,* or *The Divine Husbandman's*

Classic of Chinese Materia Medica, where kudzu is noted to relieve fever, diarrhea, and vomiting. The oldest-known Chinese written reference to the use of kudzu can be found in *Shih Ching,* or *The Classic of Poetry,* which is dated between 1000 BC and 500 BC. It has been commonly applied in a clinical fashion in East Asia, worked into herbal supplements, and incorporated into standard Chinese treatments for angina and alcoholism.

After a little more than a century in the US, the medicinal capacity of kudzu is not nearly as exercised or developed. But that's not to say people aren't exploring it. A 2011 study in the *Journal of Ethnopharmacology* found that *Pueraria lobata* reduced the blood glucose levels in diabetes-induced mice. In 2016, a *Laboratory Animal Research* paper found applications of *Pueraria lobata*—when fermented and mixed with yeast extract, dextrose, and *Lactobacillus brevis* (a species of lactic acid bacteria)—could be used to reduce inflammation in mice induced with IBD, or inflammatory bowel disease. And a 2012 study in the *Archives of Pharmacal Research* assessed the antioxidant activities of kudzu root, finding that it could positively treat and reduce inflammation. These are a series of snapshots into the varied ways researchers have evaluated the possible medicinal benefits of one abundant plant in the South. But this body of findings is still emerging and surprisingly sparse—digitized mentions of kudzu in peer-reviewed Western medical literature only go back to 1942. Modern Western understanding of successful medicinal uses of the plant on humans is minimal, and it is not used in pharmaceuticals, although it is incorporated into nutraceuticals. The extract of the plant joins a long line of healing alternatives sold in abundance by online retailers like Amazon and eBay, dispersed to millions of customers across the world. Not approved by governing agencies like the US Food and Drug Administration—which explicitly regulates dietary supplements but does not review them as drugs, but as food—these over-the-counter solutions are not nearly as commercialized as their doctor-approved counterparts. They're also not as safe to consume. In 2021, Christie Aschwanden at the *Washington Post* reported on over a decade of cases of dietary supplements including "prohibited, unlisted and even dangerous ingredients." So where does kudzu, with its millennia-old legacy of success in Eastern medicine, fall within these ranks?

Dating back to the sixteenth-century days when Queen Elizabeth I reigned over England and Ireland, ancient accounts of kudzu's healing properties were what initially captivated Scott Lukas. A director of the Harvard-affiliated Mc-Lean Imaging Center, director of the McLean Sleep Diagnostic and Treatment Center, and a professor of psychiatry at Harvard Medical School, he has been researching substance abuse since 1975. From the onset of his career, Lukas has been interested in finding alternative ways of treating substance use disorders. In 1995, he launched his own lab; operating out of McLean Hospital and Harvard Medical School, the Behavioral and Psychopharmacology Research Laboratory specializes in brain research in substance abuse and neuroimaging. His intention was to branch off and do more alternative medicine and different types of nutritional supplements, what he calls "off-the-beaten-path" medications. That fixation is what first led him to the vine spotted all over the South's many landscapes.

"I was doing some reading and came across this article on hamsters that were being treated with this concoction, a Chinese herb, of which kudzu was one of the ingredients . . . and I thought, 'Oh my God, that's really intriguing,'" Lukas recalled. The doctor had stumbled upon a 1993 study in *Proceedings of the National Academy of Sciences of the United States of America* that found alcohol consumption in a Syrian golden hamster was significantly reduced when it was given a compound containing *Pueraria lobata*—otherwise known as kudzu. An idea began to form that perhaps the results could be replicated in a human model. He reached out to the principal investigator behind that study, who reportedly wasn't open to collaborating with him. That didn't deter Lukas; he took matters into his own hands. "I did some more research on my own, found a whole bunch of articles in Chinese, that I paid somebody out of pocket to interpret for me," he said. That was his first introduction to kudzu—one that led to a decades-long relationship.

His hunt continued, during which he quickly discovered that tracking down research originally published thousands of miles across the globe was a taxing job. "That was one of the hardest things, getting my hands on the Chinese-only manuscripts," he said. This was the early 90s, when dial-up computers were still new to the scene and filing cabinets were a prime storage option. Lukas was up against a lack of proximity, a lack of literacy in the Chinese language, and a lack of well-developed literature search engines. Google Scholar wasn't established until 2004. "A lot of the studies I couldn't tell, because it's all in Chinese, for God's sake," Lukas said. "A lot of the time they had an English abstract, and so I could get a sense of which one I wanted. So I had to be a little bit judicious. But some of them would just be anecdotal reports. Sometimes they were real studies. . . It was an interesting process," he observed.

Kudzu is marketed in China as a purified drug that's usually given as an injection, typically in the form of ampules or a single-dose prescription in liquid form. "In China, they have them in the emergency room for when a patient comes in who is bleeding profusely and they need to increase blood flow to the brain and to the heart," he said. In Japan, he notes it's "actually used to treat the alcohol withdrawal symptoms as well, and heavy-duty signs and symptoms of the flu." Scott claims that in East Asia, it's commonly deployed to help express alcohol out of a system, along with a range of applications for different ailments. The doctor describes the way it's said to work as follows: kudzu mobilizes the alcohol out of your gut, into the bloodstream, where it's now able to be metabolized before making its way into the urine. "It helps cleanse the body a little bit," he added.

Although digging into Chinese research on kudzu proved to be challenging, it wasn't impossible; Lukas quickly found a pattern in the body of Eastern medicinal research he was reviewing. "The one common denominator was this kudzu," he recalled. In most cases, it was applied in combination with other ingredients. He saw an opportunity to challenge that. In order to do so, the researcher would need to develop a new standardized extract of kudzu root that could be applied to humans, which could then be investigated to see if one plant with an ancient past could really reduce alcohol consumption—not just in hamsters, but in people. At this point, Lukas says he ordered dried kudzu online from one of the "bigger" companies that sells it—who, he refuses to name—and ran his first experimental study. "The study was completely negative, and I almost gave up," he said. "I almost walked away from that thinking, 'Oh my God, this is—there's nothing to this. It's all hearsay, it's all anecdotal reports. Nothing." Lukas would later ask someone he worked with to analyze the ingredients of all the other items he had purchased from the kudzu product line. What they found was shocking: there was absolutely no kudzu in it.

Discouraged and despairing, Lukas did not know that things were about to change—thanks to an introduction to David Lee, a chemist who was struggling to find funding because he lacked clinical expertise. Lukas, who already boasted years of clinical expertise, could offer a solution that benefited them both, as he needed someone to make the extract he wanted to test. One thing led to another, and suddenly the two were working together to create a bedazzling new product made up of various ingredients derived from raw kudzu. Just think—a bio-organic chemist and a pharmacologist, teaming up to harness the potential of one emblematic plant. "It was just like a match made in heaven," Lukas said. Lee set out to make the kudzu extract Lukas was looking for, using kudzu imported from China. Lukas tells me the dried root of kudzu found in the southern US has fewer active ingredients than the type that flourishes in China. "The amount of the active ingredients in the kudzu here in the States is useless to me. It is very, very low," he said. His theory on this has something to do with the variant they like to incorporate, and how a plant developed in its natural ecosystems has higher concentrations of natural insecticides—but it's purely anecdotal. Before long, the researcher had what he needed to test his hypothesis. Dosage perfected, it was time to test this out with real people. For that very first experiment, he was prepared for the worst. "There's a risk of dying. It's that dangerous," he declared, evoking a hint of cinematic drama balanced with a somber, matter-of-fact tone.

With specialists on standby, the team assembled for the clinical trial with no concrete idea if there would be side effects for the extract they developed. And with such questions on the line, the purpose of the first study was safety, above anything else. "We're all sitting around, and it's like you could hear a pin drop," he said. Lukas remembers the doctors on call, prepared to act if anything were to go awry. "I still remember this . . . I got two physicians on either side of me, the guy had already consumed the kudzu pills, and we'd already told him that there's a potential here. We got doctors on board, I had a code cart ready to go," he said. That single participant then started drinking . . . and nothing happened. "It was like, 'Oh my God,' and you could just feel the sigh of relief from everybody," Lukas recalled. The doctor characterizes that first experiment as "stunning." "[It] worked beautifully," he added. That's when he got the green light to test this in a laboratory setting. In order to optimize this research into kudzu's impacts on alcohol, the doctor and his staff decided to create a lab that felt as homelike as possible. Working out of a room in Harvard's McLean Hospital, Lukas and his team created an ecosystem intended to relax their study participants. They outfitted a hospital room with everything from a cozy reclining chair to plants, adding a splash of greenery.

It's easy to imagine from the perspective of the trial subjects: Open the room's single door, and you step over a threshold into a pseudo-studio apartment. The whitewashed walls, fluorescent lighting, and sterile atmosphere that accompany most hospital settings disappear beyond that nearly Narnia-esque doorway. As you move into the space, you're told that the refrigerator is stocked with plenty of water and your preferred brands of beer. But what you don't know is that lurking in that test ecosystem is an innocuous table with a hidden scale mounted inside it. Lukas had previously cut out a section from the top and superimposed a ceramic tile to shield the mechanism within— "everybody was upset that I'm ruining this beautiful cherry table"—which is precisely how he and his team are able to monitor how much alcohol is in participants' systems and at what rate it is consumed. That clandestine piece of furniture was purposefully placed next to a reclining chair, giving the illusion of comfort. The subjects were told they must always keep the frozen glass mugs they're drinking out on the table or risk violating a hospital safety policy. Unseen by the naked eye, a slew of discreet video cameras are littered throughout the room, one disguised in a corner within a compact tree, another housed in other unassuming plants. A few others are intentionally made visible to the subjects in the room. On the other end of these sets of lenses are Lukas and his team, keeping track of the participants 24/7. Like spies on a secret mission, they are intently monitoring subject behavior for the duration of the experiment.

Over the course of one week, all participants were treated with kudzu pills or placebos. The team testing their reactions first sought to get a baseline of how much the human subjects tended to drink over a two-hour period. From there, Lukas repeated the same experiment, except this time, they only provided subjects with placebo pills for the testing. All the while, the people being scrutinized had no idea they were being measured sip for sip. The mugs conscientiously placed back on the table served as quantitative tools or behind-the-scenes insights into human behavior for an eager team of medical researchers. Back in the hidden observation room, they were using that frequency to measure exactly when and how much alcohol had been consumed. Both date and time stamped, the data helped the researchers confirm when there was no more kudzu or placebo in a participant's system. Once again, the experiment was repeated; the team swapping placebo pills for kudzu pills and vice versa. And so these experiments starring the "vine that devoured the South" carried on, growing in notoriety like a congregation of unbridled kudzu.

What eventually emerged was an intriguing, if not altogether unsurprising, pattern. According to Lukas, it became apparent that kudzu "hands-down" reduces alcohol consumption. The results unveiled a common theme, as participants taking the kudzu extract would consistently increase the time period between each drink—meaning the kudzu extract effectively tapered down any desire for binge drinking. Not only did it appear to delay an impulse for the next drink, but it did that without blocking consumption outright. And when it came to chronicling any adverse side effects of the kudzu extract, the slew of experiments revealed something the researcher welcomed—there were none. "Number one, it lowers cholesterol. Number two, it lowers triglycerides," Lukas said of the observed, positive impacts.

Theories abound as to why. A possibility Lukas cited has to do with all the ingredients in the kudzu extract they've patented being water-soluble, which theoretically could mean they rapidly cleared the body. His experiments have led him to believe that it could also be possible that following a first dose of alcohol, blood alcohol and brain alcohol concentrations may elevate more quickly—when consumed with the right cocktail of kudzu. According to Lukas, this is because components of the weed can redirect blood flow. "It's turning off the desire for subsequent drinks because it's so effective in bringing the alcohol to the brain. Now at the same time, it's not impairing them more," Lukas said. Another of the main takeaways from his research so far is that kudzu can help minimize the amount of booze a person wants or needs to drink. "If you want to be mercenary about it, it's cheaper. They can get just as high, they're feeling okay after half the amount of alcohol that they would normally drink. So now there's money in their pocket. They don't have to drink as much. Their blood alcohol is lower and [they're] less likely to drink and drive," he observed. "That's, in a nutshell, what the kudzu story's about."

A conflict-of-interest statement has been flagged within several papers summarizing this body of work. One example is a 2015 study that Lukas coauthored in *Drug and Alcohol Dependence,* which disclosed the following: "Drs. Lukas and Lee hold a patent for kudzu extract to treat alcohol abuse and dependence. McLean Hospital has licensed the production of kudzu extract (NPI-031) to Natural Pharmacia International (NPI), Inc., which markets it as Alkontrol-Herbal™. Dr. Lee has a financial interest in NPI, Inc. Other authors have no conflicts of interest to declare." An earlier study, first published online in 2012, and in print in 2013 in *Psychopharmacology,* noted a similar conflict of interest statement but also specifies that "Based on data presented in this study, Drs. Lukas and Lee applied for, and were granted, a patent for kudzu extract to treat alcohol abuse and dependence."

Meanwhile, Lukas still has questions about kudzu he would like to explore. A few years back, he failed to get funding for a kudzu-themed study looking at the brain concentrations of alcohol, which forced him to hit pause on those plans. "There's a lot of evidence in the Chinese literature that this *pueraria* increases blood flow to the brain and to the heart. Deep down, I really believe that's the mechanism. But the scientist in me says, 'But you really got to show it,'" he expressed. Lukas wants the chance to do exactly that, noting it would be "the definitive nail in the coffin." The lack of peer-reviewed research in English on kudzu's medicinal applications has certainly complicated things. If you ask Lukas, the absence of clinical research on kudzu comes down to an industry-wide hesitancy to delve into nutraceuticals, which are not regulated by the FDA. "My first knee-jerk reaction is, yeah, why not? Why isn't there more on this? And then it hit me," he said. "This is not *unique* to kudzu. There's so much about nutraceuticals in Chinese medicine, Chinese herbal medicine, Japanese herbal medicine, from Indian cultures, and there's a plethora of different drugs that are used in Southeast Asia, in the Aborigines. There's all these things out there, they just don't have the studies."

Another considerable roadblock he identifies is the process of acquiring a high-quality product to work with. Too often, researchers don't know the precise source of the nutraceutical products they are testing. The problem lies with uncertainty around quality and the specific content of the material. And many don't have the resources needed to help vet that. "When people don't do that, my eyebrows go up. That's why this research has such a torrid, torrid history," he said, emphasizing the need to provide the source of the nutraceutical that demonstrates the material has been tested, it meets quality control standards, and the author knows exactly what it's made out of. "That's why there are anecdotal reports that are not worth printing. Because we don't even know what's in them. And that's the danger of this field."

Clinically testing kudzu might be novel in the US, but it's old news in China and Japan where the weed is native. Two Chinese websites, cnki.net and wanfangdata.com.cn, are comparable to *JSTOR,* a leading digital source for academic journals and peer-reviewed papers in English. Kudzu is referred to as *pueraria* in most Chinese academic studies; however, in Chinese, kudzu is 葛 (*ge*), which doubles as a surname. The most searched or related search items on these two research hubs have to do with 葛根 or kudzu root. A 2022 search for "kudzu root" on these websites uncovered 19,413 and 47,394 results, respectively.[1] When using other variations of keywords related to kudzu, like "kudzu leaf," the number is significantly less.

The body of digitized research in Japanese tells a similar tale. When searching on *J-STAGE,* a journal platform for Japanese academic journals, using the Japanese letters for *kuzu*—or クズ—8,570 studies with *kuzu* in their title appear in the research portal. Comparatively, searching for peer-reviewed papers in English using a digital search engine like Google Scholar produces around 1,600 hits. Of those, 790 include the search term "clinical," replicating the usage of the vine in alternative Eastern medicine. Roughly 98% of publications in science are produced in English, according to a 2020 paper in *PLOS One*—further underscoring the scientific shortfalls concerning one of the most eminent plants in the southern US.

While these numbers may only scratch the surface, what they indicate is that although kudzu isn't extensively studied in Western medicine, the same can't be said for other parts of the world. As far as Lukas is concerned, the American indifference in the medicinal utility of the vine thus far presents an untapped opportunity.

"It's just not as highly publicized," he said. "That's why the lay public hasn't gravitated to it."

1. These results are based on keyword searches in Chinese and Japanese that included "kudzu" in study titles. They are reflective of individual papers.

9

GOT GOATS?

A goat wanders aimlessly onto an open field in a pocket of an expansive park flanking a thicketed mountain terrain. It is content, as well-fed goats tend to be. Sounds of unhurried chewing accompany its steady, leisurely path. Verdant leaves jut out of the animal's moving jaw, a few pieces falling to the ground in a wistful arc, resolutely ignored in the pursuit of lunch. If you look closely, you will see that the goat isn't just devouring any old patch of herbage. With the quiet confidence of a creature having all the time in the world, it is working its way through a stubborn patch of kudzu, unknowingly keeping the egregiously large plant from taking over the soil beneath its hooves. The goat knows no better. All it can focus on is this bountiful, gratifying supply of food.

Qinfeng Guo can tell you all about the legacy of entanglement connecting goats with kudzu. A research ecologist at the USDA Forest Service Southern Research Station, which operates out of Asheville, North Carolina, the fifty-seven-year-old researcher originally hails from Chengde, a city in northeast China—where kudzu is not a monstrous foe, but a cherished friend. In fact, Guo says in China, people intentionally plant kudzu, simply because the vine is in such high demand. For more than two decades, he has studied invasive species in the US. Kudzu is one he is professionally aware of but personally very familiar with. What makes the weed a bane in the United States and a boon in China comes down to how it's used—and not used. In China, people have historically faced high proportions of poverty, which the scientist believes is why kudzu's utility in Eastern medicine, cooking, clothing, and material production has persisted for a millennium. It's because of that ancestry that he "loves" framing the plant through the lens of the past. "Here in the US, we have so many resources, we don't need kudzu," said Guo. "But in Chinese history, kudzu has always been part of human life."

In 2011, Guo coauthored a paper in *Biological Invasions* analyzing the human dimensions of the invasive vine and the impacts of natural "enemies" on its range. "You have this species that will climb all the way to the top and kill the native plants, native trees, and then the forest is going to shrink from the edge," he explained. That species is, of course, kudzu. Forest edges and disturbed habitats are the spaces where it predominantly dwells, which tend to be overlooked in terms of importance compared to pristine forests. According to Guo, dismissing its scope of impact because of where it thrives is the wrong way to look at it. "The edge of a forest often holds a lot of rare species, not only plant species but also insects, bird species. The forest edge and other kinds of habitats, grassland or abandoned fields, crop plants, that's where kudzu really lives. But those areas are also a unique habitat for many native species. That's our major concern. That's why we worry about this species," he noted. "Kudzu is still a big problem."

More than 350 species—285 of them insects—were surveyed by Guo and his colleagues in that 2011 study. Lead author Zhenyu Li was based in Beijing, where they conducted their fieldwork. The insects under investigation were reviewed as potential forms of biological control agents that could eradicate kudzu, alongside fungi, bacteria, nematodes (roundworms), and land mammals—but none proved to be up to the task. His paper refers to what Guo describes as the "many" researchers who have tried to go to places where it is native—like China and Japan, where there is an abundance of kudzu and yet its growth is under control—in order to find an effective biological control agent. "They all failed. I mean, they were not successful. Even in Asia," he said. Guo and his team saw similar results; kudzu's natural enemies seemed to have little consequence to the vine in the wild. "We cannot find a single insect or an enemy for kudzu," Guo noted, before concluding that human use is the largest, most effective management tool in kudzu's native range across East Asia. According to the USDA researcher, if Americans started maximizing their usage of invasive species like kudzu, the results would be just as widespread. "That impact would be huge," he said.

There are no peer-reviewed papers in English, Japanese, or Chinese that do more than estimate how much land kudzu covers in its native range. A

2010 paper published on *J-STAGE*, クズ (*Pueraria lobata Ohwi*), estimates that the plant covers all of Japan; from Hokkaidō to Kyushu, including Okinawa. That paper describes kudzu as a "valuable Chinese medicine" and "one of the most expensive starches in Japan." *Baidu Baike*, or China's version of Wikipedia, also suggests that swaths of the vine cover the entire country with the exception of a few provinces, like Xinjiang and Tibet.[1] "Some people [in China] use kudzu as plant capital if they have land. Others use it for medicine, for weaving, for the skin, and for clothes. Every part of the plant can be used," said Guo. "But here in the US, nobody is doing that."

What a few Americans are doing is delving into the potential of kudzu in helping power a greener future. Repurposing the weed doesn't need to be limited to certain industries where it has been popularized in the past. According to Guo, if Americans were to start utilizing kudzu for biofuel or for livestock grazing, goats and all, that would not only be beneficial to the country but would help us utilize sustainable resources. It's the case of resourcing something that needs to be contained and building up a new, clean source of energy while also mitigating and controlling kudzu expansion. "The Biden administration is really trying to develop green energy," noted Guo, "and kudzu and some other invasive plants that can be used as biofuel could be a big potential source for energy." He warns this would need to be done in a "very cautious way," so as to avoid any possible missteps that come with economically incentivizing an invasive plant. "If, after some careful studies, the US government decides, 'Wow, this is a potentially big biofuel candidate. Let's use it, anywhere we can find it,' maybe we have to plant some," Guo said.

Biofuel à la kudzu is a beguiling idea, but the concept of using plants to power our grid is nothing new. The potential of kudzu to be used for biofuel has been entertained by the US government as far back as 1979, when NASA experimented on it to test its capacity. Critics of the push to use invasives for energy conversion argue that bioethanol production is both complicated and costly, and establishing a system for biofuel that depends on a plant en route to eradication makes little sense. In response to this, Guo stresses that stringent prerequisites would be vital if the weed, or any other invasive species, is

1. Remote and predominantly a Buddhist territory, Tibet is governed as an autonomous region of China. "Tibet Profile," *BBC News*, April 26, 2019, https://www.bbc.com/news/world-asia-pacific-16689779.

to be considered as a fuel for the future. Something he's thinking about as a possible avenue to get to that point would be genetically modifying kudzu—manipulating the plant to grow less rapidly and become unable to produce seeds. At present, it is nothing more than a far-fetched idea. "We don't have enough funding. We don't have enough research now to know the potential for kudzu to be a biofuel plant," he said, citing the need for increased understanding of the possible manipulation of kudzu's genetic makeup before this can be considered in the field. "So that's the reason we don't know."

Livestock grazing of kudzu is a less controversial control method that's been implemented across the world. In its native habitats in China and Japan, kudzu is commonly used in livestock grazing, which Guo describes as an important factor in successful kudzu management. In the United States, however, there are legal limitations to this. Although more than 1.9 million "cattle equivalents" graze on public lands, as first reported by the *Revelator* in 2019, federal agencies must authorize such grazing. Under current law, ranchers lease grass for grazing but hold no "rights" to the public land itself. According to Guo, these conditions get in the way of a plausible win-win opportunity. But just like regulating kudzu's growth and tracking its spread, in the US, the solutions are complicated. "In China, that's not a question," he said, because the concept of private landowning doesn't really exist. If the Chinese government directs ranchers to graze in areas where kudzu is abundant, "they follow," according to Guo. "You have to have a government that can say, 'Yeah, let's do some more work with kudzu.'"

The use of herbivores by scientists, federal agencies, municipalities, and the private sector to manage woody and other invasive plants has proven to be a common form of kudzu control. If the grazing is continuous, pigs and goats have been known to eradicate kudzu from entire fields. According to a 1991 report by the USDA Southern Research Station, kudzu leaves provide around 15–18% crude protein (CP) and are quite palatable to livestock. And a 2019 article in *Agriculture* assessed the vine's nutrient composition, degradability, and palatability to find that it would make a "valuable potential feedstock" in the United States. It referenced a 1947 federal experiment conducted by the Soil Conservation Service in Puerto Rico when the weed was used as a source of livestock grazing for a herd of Guernsey cows as well as oxen and goats.

Ever since, much has been written about goats and kudzu—protagonists and antagonists, heroes and villains. Theirs is a symbiotic relationship, one where a deliciously despised plant is taken down by a quirky, beloved animal, a tale that humans devour with the same gleeful appetite as a goat let loose in a tangle of kudzu. Over the last few years, a flurry of articles on the creatures being used to curb the South's notorious weed have hit newsstands. Outlets like *BBC News,* the *Washington Post,* and *National Geographic* have all covered the phenomenon—as have local news organizations across the region, reporting on community-based efforts to tap into this common modus operandi to get rid of unwanted kudzu.

One of those is the *Berkeley Independent,* a newspaper in Summerville, South Carolina, which has covered the work of researchers at Clemson University who are evaluating the effectiveness of goats used as a form of invasive species management. First launched in 2014, that project has since wrapped, with preliminary results suggesting that the grazing animals have significantly reduced cover of kudzu, alongside a slew of other invasives, including Chinese privet and Japanese honeysuckle. Goats making major dents in the thickets of plant biomass allowed for more unfettered access for those working to eradicate the invasive populations, to spray herbicides, and to get rid of leftover vegetation. However, introducing animals to landscapes with creeks and cliff slopes—like the site by South Carolina's Hunnicutt Creek that this project ran on—can usher in water pollution and sediment buildup to an area. When goats are involved, active management is also a necessity. The Clemson University scientists relied on electric fences to keep their herd in check during their invasive removal à la goat experiment.

Hundreds of miles away is Red Mountain Park—a nonprofit site spanning 1,500 acres in Birmingham, Alabama, where kudzu used to be a colossal headache. In 2018, Red Mountain Park director of natural resources Rachel Ahrnson told the *Alabama News Center* that the staff got creative with an overabundance of kudzu and privet on their land—which had been neglected for fifty years after nearly a century of being an iron mine—by renting goats. A herd of one of the world's oldest domesticated species chomped on the unwelcome plants spread across 250 acres of the park for a year. When it was over, nearly two thousand volunteers, "armed with loppers and hedge trimmers," cut back the plants to curb any regrowth. And yet, goats aren't the only animals to feast on the southern vine of abundance. They are merely one of many, joining sheep, alpacas, cows, llamas, deer, horses—all types of grazing animals feed on the coiling plant.

Livestock can cause lasting damage to woodlands, according to a 2018 study published in *Agriculture, Ecosystems & Environment,* with soil and vegetation degradation as the fallout. Comparing grazed to not-grazed woodlots in Wisconsin, the researchers conducted vegetation surveys and assessed soil properties, concluding that unfettered cattle grazing produced generally negative ecological consequences. This is only one of the latest pieces of evidence in a body of literature that finds a sizable scope of impacts—endangering wildlife, contaminating water quality, bringing about soil erosion, and wreaking havoc on native vegetation—that compounds with cattle grazing's contribution to climate change, according to a 2016 *Grist* news article by journalist Ben Adler.

Despite the downsides, livestock grazing is a typical go-to for ranchers and farmers looking to curtail unwanted plants. A 2021 study published in *Restoration Ecology* found that prescribed grazing by goats on invasive species in a mixed-hardwood forest reduced the cover and height of invasive shrub species, but the practice needed to be followed up by human control, like targeted herbicide, to ensure no future regeneration. The authors pointedly did not leave the animals in the woodlands for the entirety of the growing season but removed goat populations after all green leaves belonging to invasive and non-native shrub species were consumed, according to a Purdue University College of Agriculture summary of their findings. Although the field experiment proved to be a smashing success, the researchers warn that the industry lacks options for "herds for hire," which makes the treatment method thorny to deploy.

No matter, as goats are still being corralled to snack on plants people want gone in pockets of the US today. In Burlington, Wisconsin, a herd of goats is actively being used for land management, with the furry mammals minimizing populations of invasive honeysuckle and non-native buckthorn, alongside other undesirable vegetation—like kudzu's nearly equally illustrious twin, poison ivy. Chloe Lewis at the *Highland Echo,* a student paper in Maryville, Tennessee, published an article in the fall of 2021 chronicling how local organizations introduced goats to graze the kudzu growing in the Maryville College Woods. Over in North Carolina, Madison Elliot at *Spectrum News 1 Charlotte* reported that the city of Belmont hired a "goat whisperer" in the spring of 2022, as well as his roughly two dozen goats, to take on kudzu growing on public lands.

But not all are eager to explore the possible multipronged potential of unleashing an army of goats to both eradicate the "vine that ate the South" and use the uninvited growth as a source of cheap food to satiate hungry livestock. Stephen Enloe, a professor at the University of Florida's Institute of Food and Agricultural Sciences Center for Aquatic and Invasive Plants, thinks exploring large-scale uses for invasives, like kudzu used for livestock grazing or bioenergy, should be approached very cautiously. "The use of invasive plants for alternative purposes like this, it opens a whole new can of worms," Enloe advised. "Every time people take economic interest in a species, then the game immediately changes. It goes from being something that may have been a novelty, to something that is looking to go into mass production to make money off of."

It's a vicious cycle, one with a tendency to repeat itself. The difficulty lies with creating supply and demand from something that can be a wrecking ball, striking native ecosystems by disrupting biodiversity. "It has a long history of going astray," he explained. The researcher emphasizes how this has been done in the past with other invasives; emperor's tree, Barbados nut, and Chinese tallow tree are prime examples. In the 1900s, the USDA promoted the tallow tree as an oil crop, widely distributing it in the hopes it would establish a soap-making industry, which quickly led the tree to outcompete native plants and invade forests and coastal habitats across the Southeast. Today, the tallow tree impacts an estimated 185,000 hectares in southern forests, with the cost of control projections specific to the forest industry ranging between $200 and $400 million over the next two decades.

"Those economies never materialized, and we were left with vast areas of these very aggressive species that rapidly became invasive," said Enloe. It's not that he doesn't support finding alternative uses to invasive species—he's just skeptical about moving forward with any solution that sets the stage for mass production. "We have too many examples. We've been burned too many times with plants that people have promised the world. And nothing ever materialized out of it except a problem."

10

MISSING IN ACTION

Before me is a shrouded trail, emblazoned by moss-draped cypress trees. It appears gilded in sunset, drops of mist lingering on leaves, whimsical remnants of a light dusting of Asheville's afternoon rain. Between these otherworldly scenes, I feel free, trekking onward as my scruffy-haired guide waves his palm, gesturing wildly as he chatters on about the many uses of fungi. A walking mushroom encyclopedia, if I've ever seen one. I settle for these little moments that provide a promise of peace. But that sensation of serenity doesn't last. Merely a day later, I find myself up to my ankles in some type of prickly bush as I wade through an open field outside the city limits of the charming North Carolina hub and its slew of bookstores, a type of turf I'm far more familiar with. My limbs heavy with the imminent knowledge that there are all kinds of critters bustling about their day-to-day lives around me, I feel the frown already etched into my face miraculously manage to further deepen as lines of distress and general discontent carve a map on the open plain that is my forehead. If I could, I'd snap my fingers and be back inside my rented room, curled up underneath the softest of blankets, utterly unaffected by the elements outside. Alas, that power is not mine to wield when I want. I am but an unimpressively average human who can no more protect herself from the great outdoors than an albacore can take shelter from a hunting hawk. Feeling far closer to prey than predator in this open expanse of field, I head in the direction of my destination: more than an acre of open farmland, corners swaddled in kudzu like newborns cloaked in quilted blankets. Here is where

I will pick handfuls of the plant, to be stir-fried, ground down to a starch, or made into a tea.

And for the next hour, I tug leaves off of seemingly infinite patches of greenery, stuffing my hard-won prizes into a tote bag hanging at my side. I am starting to adjust to the environment I occupy, finding an unforeseen sense of satisfaction in the task I've committed to do. Perhaps I *can* do this after all, I think. Perhaps this is the path I've always been meant to find, I wonder. For the longest of moments, I am suspended in that illusion of belonging, of finally finding a place free of hooded, heavy stares, free of the invisible cloak of persecution, knitted together by a past I can't seem to outrun. Perhaps here I can belong. It's a thought I dare to entertain for the shortest of seconds. At least until the smallest of somethings, an indistinguishable insect fluttering about, lightly nips at my exposed ankle, startling me so deeply that my bag of foraged goods goes soaring through the air, blossoms scattering in the sky like a flock of violet-backed starlings jumping into flight. With that, my desire to stay any longer begins to splinter, my remaining courage shattering like specks of sand impacting minuscule particles in a bed of soil. If we were keeping score, it'd be an easy point for the other side: *The Great Outdoors 1, Ayurella 0.*

I grew up in urban expanses, where green spaces were only found in manicured parks miles away from the concrete units I called home, and the typical tree lining nearby roadways was shaped like a fan. Where sputtering air-conditioning units and fossil fuel–guzzling vehicles drowned out the sounds of nocturnal creatures. Where a neverending number of walls were erected to keep the divisive world out and you in; a self-made prison of sorts, gilded, ever-expanding cages we were forced to confine ourselves to. Where stargazing is swapped for screen watching, a tiny box with taped bunny ears fixated on by a family of five squished into a dingy living room, walls shared by neighbors who preferred hollering to talking. This was my world—a void made of mindless machines, where I was but an unassuming, unimportant cog. This was *the Big American City*: a haunted, polluted space entrenched with stamps of industrialism, everyday evidence of redlining and systemic racism, and desolate poverty a stone's throw away from sickening opulence. Where currency trumped compassion, beauty beat brilliance, and the individual always came before the collective. Mine was a world where flora and fauna were polyester or plastic more often than not. Where that which was wild was conquered, broken down, and kept under lock and key. Where the unfamiliar was bad, and so was anyone who resembled anything new. Where outsiders—be they pesky critters or children with different shades of skin—weren't welcome.

Enclosed within slabs of brick, in houses where laundry spaces doubled as bedrooms, I dutifully read fanciful tales about the great outdoors, fables of adventure and glory, starring predators and prey and impossible tasks to overcome. I plucked voyages straight from their spines and let them swim in my daydreams. Between those pages, penned by authors far more courageous than I was, I lived several lifetimes, fueled by the riches of fictional escapes set in the python-ridden Everglades or the salt-drenched Dead Sea. On the wings of the thrilling prose created by someone before me, I could pretend: slip out of my safe, sterile world into the fearless lifestyle I'd always wistfully, bitterly desired. At the cusp of youth, these make-believe characters lived in my head rent-free. When I needed to, I would wear them like indestructible armor. Their personality traits and physical characteristics would meld into my own, intangible shields that could protect me from reality. When the world around me became too heavy, I would flip open a book that would whisk me away to faraway lands. Within moments, I'd be lost in scenes where pirates with hearts of gold pillaged the rich and gave to the poor, mermaids lured bad men into unforgiving seas, and unlikely protagonists became heroes of immense cities. These were spaces of refuge, spaces where a quietly uncertain brown girl transformed into a powerfully glorious goddess overnight. Magic and mayhem ruled these make-believe visions. But that's all they were: artificial longings safely ensconced in my mind. For the longest time, books were my go-to ingredients for a priceless cure, applied when I needed to escape. An anti-venom for a day-to-day existence wrapped in sorrow, hurt, and fear. A way to freeze time and leave the tough stuff behind. Until the day they were no longer enough.

This is precisely why I'm here, swimming in a sea of relentless mosquitoes, pesky no-see-ums, and curious spiders lurking on nearby trees. So completely at odds with my surroundings, unease mounting with every passing moment. Still, there's something profound in being so far out of a comfort zone. The air is charged, the beings making up the forest biome thrumming to a beat I can't quite understand. My pulse ricochets under my skin, anxiety spreading throughout my system like a live wire. I feel warm, like any second I might explode into burning stardust, returning to the very universe I originated from. Teetering on the edge of something new, I am unbalanced, afraid, and wholly unprepared for what's ahead. Adrenaline-pounding dizziness flirting with fear—a hesitance as the path forward shatters the safety net at the heart of my carefully curated world. All this trouble. In the name of kudzu.

A cultural staple in the Southeast, the weed at the center of my inner turmoil covers an undetermined number of acres across the US. Because it dwells mostly in disturbed habitats and at the edges of forested land, the federal agencies in charge of tracking invasive species growth don't measure, or really even know, exactly how much land it covers. Or the rate at which it spreads. Up-to-date reports recording kudzu range are limited and incomplete, missing essential data. A 2001 chapter penned by Richard J. Blaustin in the book *The Great Reshuffling: Human Dimensions of Invasive Species* cites figures from 1996 detailing how kudzu covers over 2.8 million hectares in the United States, with the largest concentrations in Alabama, Georgia, and Mississippi. The chapter concludes that the vine was "actively promoted by the government as a 'wonder plant'" and expanded to cover over one million hectares by 1946 and "well over" two million hectares at the time of publication. USDA reports since have estimated that kudzu's total land coverage may be anywhere between seven million and nine million acres. However, none of these numbers are comprehensive, as they only capture kudzu's estimated range on forested land. This is significant, as it means those measurements don't account for where kudzu grows and thrives—in areas considered "non-forest." Chris Park and Michael Allaby's third edition of the *Oxford Dictionary of Environment and Conservation,* published in 2017, defines these spaces as "land that is not used for forestry or timber production, but is used for other activities such as farming, transport, industry, commerce, and housing." How quickly kudzu spreads is also unclear. Estimates of the vine's land coverage vary, from the US Forest Service's 2015 estimate of an annual increase of 2,500 acres per year to the Department of Agriculture's estimate of as much as 150,000 acres every year.

Qinfeng Guo at the US Forest Service Southern Research Station isn't able to provide a more recent, or even accurate, number quantifying the total land kudzu covers. "This is hard to determine as it depends on the scales to measure the range," Guo told me. "The smaller the scale, the smaller [the] coverage." Initially, he suggests an estimation of 9 million acres, quoting a

USDA study from 2001. When pressed, he shares a 2015 USDA webpage that shows the estimated forest-only coverage of kudzu is an estimated 209,630 acres. He confirms that this USDA estimate isn't a correct measure, nor is it comprehensive, since it only approximates kudzu's forest coverage, missing the areas where it's known to be most prevalent. "Even the Forest Inventory and Analysis data, it only covers a certain . . . very small part of the real forest," he noted. At the US Forest Service Forest Inventory and Analysis (FIA) National Program, Guo's colleagues shed some light on this rather unsatisfying response. Sonja Oswalt and Christopher Oswalt are foresters with the USDA Forest Service Southern Research Station. Over email, they explain that capturing kudzu's modern persistence and spread is a task as tangled as a swath of kudzu tap roots. "Our program only measures kudzu on forestland—and to be honest, kudzu is one of the lesser problematic species affecting forestland, specifically, compared to much more widespread honeysuckle and privet in the South. Kudzu can't persist well in the understory, so it tends to be in highly visible areas next to roads and homes, which is why it's such a high-profile species. We can provide you with acres of forestland impacted, but that won't give you any indication of how much land at the forest edge has been affected that is considered non-forest (e.g., on farmland, gullies, roadside easements, etc. where it tends to be the most voracious)," wrote the Oswalts. According to the USDA foresters, reported forestland coverage of kudzu "will be an underestimate of total acres of land infested."

The two contributed to the 2021 book *Invasive Species in Forests and Rangelands of the United States: A Comprehensive Science Synthesis for the United States Forest Sector,* where they analyzed the gaps and challenges in current efforts to assess and monitor invasive species, especially in light of the uncertain impacts of climate change on the ecosphere. That chapter details the lack of longitudinal datasets and the broad scale of existing spatial data, which they determine makes "generating simple distribution data for many species" a challenge, "especially recently identified species not yet on monitoring lists and species in non-forest systems and aquatic habitats." Another obstacle they pinpoint is the concern around the reliability of citizen science data as well as the varied approaches to inventory and monitoring invasive species and the need to standardize these among all agencies, organizations, and programs across the country. This brings us back to the question: Does this lack of data hurt any emerging moves to examine kudzu further? Without that quantitative evidence, I'm told by a smattering of scientists in this space that it becomes very difficult—sometimes even impossible—to get the funding necessary to carry out desired research.

All of these dilemmas seem to stem from an existing, jumbled system where everyone involved tackles invasive species differently. Would a facelift of the invasive plant tracking system to align with one standard, one universal tool, and one form of measurement, assigned as the responsibility of one singular agency, help solve these problems? It's a question I posed to Guo. "That's ideal," he admitted, "but it's so hard to do. Maybe that's why we haven't seen anything like that anywhere in the US." The scientist points out how even for federal lands, different agencies have different policies regarding their work overseeing invasive plants and species. "I don't know anybody powerful enough to do that kind of work," he continued. When asked who among federal, state, and local governments is responsible for tracking the growth of an invasive species like kudzu in the US, Guo is also adamant that "it really depends."

The US Forest Service, which is a branch of the USDA, works to provide education identifying invasive species—where those species are and where they're expanding. They offer funding to state and local governments for eradication efforts and are in charge of mitigating kudzu on public lands like national parks and wildlife refuges. Accountability beyond that degree is where things get murky. According to the US Forest Service scientist, the federal, state, and local government division is complicated—particularly when you consider how much of kudzu-covered land is privately owned. "It is really hard for the government to make a policy for private and state landowners, like, 'You have to do this' or 'You have to do that.' It's not easy," he added. This gray area of accountability lends itself to something of a patchwork approach where federal, regional, state, and local agencies are left to their own various preferred methods when tasked with measuring and monitoring the growth rate and total land coverage of kudzu. It also feeds into this vicious cycle fueling a world of research where funding is limited and data is critical to demonstrating a need for funding. And it could be why so many questions remain unanswered about a weed with a reputation as muddied as North Carolina's bottomland swamps.

Exactly who is responsible for invasive species management is another complex question. A 2015 Congressional Research Service report asks, "Whose responsibility is it to ensure economic integrity and ecological stability in response to the actual or potential impacts of invasive species, and at what cost?" With responsibilities divided between local districts, states, and the federal government, the answer is something of a tangled, piecemeal approach. When it comes to measuring and mitigating the spread of kudzu in the Southeast, more than a handful of agencies are involved. The USDA and US Forest Service both play a small yet somewhat significant role in nationwide invasive species management on the part of the federal government—but beyond the Forest Inventory Analysis National Program's incomplete estimate on kudzu coverage, no one is in charge of tracking total land coverage of invasives like it. And accountability in management and eradication of kudzu dredges up a whole other set of hurdles, as the system in place lacks federal regulation or support. "It is so localized," confirmed Guo. The bulk of the responsibility for eradicating kudzu expansion lies with state and local governments; so in a state like Alabama, where kudzu has historically been a source of trouble and still persists as one, it's up to state and local officials to deploy control strategies to mitigate it. But some of those agencies are better at this than others, and some states' invasive research budgets are so strapped they ignore anything that isn't generating a lot of publicity. "We have a national map that is a view of kudzu, but sometimes you see a big hole in the Southeast. So you think 'Blank. Kudzu. Nothing,' but actually, that's not the case. It's because the state didn't have enough money to go out to measure it," Guo said. My next question circled back to federal authority: Would a federal policy implementing a set of generalized invasive plant tracking standards and reporting expectations, as well as a set of national guidelines determining these gray areas of accountability, be useful? As a federal employee, Guo is careful with his response. "What I've heard from my supervisors, my unit is, 'Well, as a scientist, you can make suggestions, you can express your opinion based on the data, but we should never tell anybody to do anything,'" he affirmed.

Once, there was a regional organization that tried to fix some of these glaring problems. Founded in 1999 with the purpose of facilitating invasive plant research and communication across the South, the Southeast Exotic Pest Plant Council (SE-EPPC) dissolved in 2018 after launching eight chapters in Alabama, Florida, Georgia, Kentucky, Mississippi, North Carolina, South Carolina, and Tennessee. The official statement as to why the organization disbanded is that it accomplished what it set out to do. University of Florida professor and invasive plant extension specialist Stephen Enloe was the last standing president of the SE-EPPC. He tells a somewhat different story, one that translates to a lack of resources. In a tale as old as time, one of the primary reasons the council was discontinued came down to funding. "A number of things happened," said Enloe, "one being a lot of budgets got hit through the different budget crunches that we've had." The former SE-EPPC president says the impact of the conference and bringing people together "began to be diminished" because there were fewer and fewer members who had the budgets to actually travel. "It comes back to that funding issue again. The organization basically reached a point where we had to take a hard look at ourselves and say, 'Have we accomplished what we set out to do?' And we looked at the state, and the supplying councils that were functioning and thriving across the Southeast, and we had to say 'yes.'" Other former members of the agency echo Enloe's sentiment: SE-EPPC disbanded primarily because the individual state chapters didn't really need an umbrella organization to help them function any longer. In spite of that, it was considered a helpful instrument for maintaining greater momentum at both the state and regional levels, but a drop in engagement within several chapters that was spurred by resource scarcity was what contributed to the demise of the overall organization.

SE-EPPC was formed because of an overwhelming lack of disconnect across the invasive species research space, according to Enloe. A fitting example is the way various local, state, and federal agencies use different methods and measurements to monitor invasive species growth. "Communication on these issues historically has been a major issue because a lot of agencies just really weren't talking to each other," he continued. SE-EPPC was founded to address that by fostering the establishment and growth of state invasive plant councils across the southeastern US. In its prime, it saw stakeholders spanning local, state, and federal entities beginning to work on coordinating management efforts that would have more meaningful impact over larger scales. At a micro level, Enloe compares it to a typical landowner stumbling block with invasive species management: You'll never fully get rid of a pesky weed that ignores boundary lines if those around you lack the same drive or are bulldozing ahead with a conflicting strategy. "You can work your tail off on a property, but if your neighbor does nothing, then that problem is always going to be creeping back over your fence line, or spreading rapidly," he noted. One of those plants is, predictably, kudzu. "Kudzu has a long and circuitous history in this country. I like to say it's a plant with really good intentions with unfortunate outcomes," said Enloe. "It's just that kudzu was so charismatic. And within the world of invasive species, charismatic often gets the attention."

Something has become increasingly clear—when it comes to our collective invasive species understanding, it all boils down to data. An abundance of quantitative data both educates and informs decision-making at the national, state, and local levels, helping determine what research gets funded, what questions are explored, and what potential expansion of information is ultimately cast aside. It is why I am baffled by the limited nature of quantitative data on kudzu is in the US—and how laborious it is to access. How much forestland and non-forestland is covered by kudzu in the US, how much money has been invested in mitigating kudzu populations, and when it comes to tracking the growth, economic impact, and management of kudzu, who is responsible for what? Working this out seems straightforward enough. It is anything but.

My quest for answers begins in Georgia, home to the Early Distribution and Detection Mapping System, or EDDMapS, a database powered by voluntary reporting of citizen scientists and by the Forest Inventory and Analysis Data. After reaching out to the state's department of natural resources, I'm contacted over email by Lynne Womack, a forest health coordinator at the Georgia Forestry Commission, which, according to its website, works to "promote, protect and conserve healthy, sustainable forests." I learn from Womack that the team relies on data published by the federal Forest Inventory and Analysis Unit, which estimates that kudzu is currently found on approximately 32,850 acres of forested land in Georgia. The unit doesn't measure kudzu's presence in the nonforested areas where the vine is most commonly found. The only species that the state agency follows for exact numbers is cogongrass. Other than that, tracking invasive plants in Georgia is also done through EDDMapS, which Womack stresses is not an exact amount of acreage. The numbers are difficult to determine because in Georgia, approximately 91% of forestland is privately owned, and that information may or may not be available as a public resource. As for economic investment, the Georgia Forestry Commission isn't currently spending any money on controlling kudzu. Womack clarifies that her agency controls a small amount of land and that kudzu is not included on the list of species in the agency's landowner cost-share program. At that point, she directs me to other agencies that control more land, such as the Georgia Department of Natural Resources, Georgia Department of Transportation, and US Forest Service, all who she suggests may be spending money on mitigating kudzu.

West of Georgia's state line is Alabama, where tracking and mitigating kudzu populations does not fall under the purview of the Department of Agriculture and Natural Resources. After reaching out to the agency, I am directed to the Alabama Cooperative Extension System, which is headquartered at Auburn University. From there, I am put in touch with Nancy Loewenstein, an extension specialist in Forestry, Wildlife and Natural Resources at Auburn, with whom I have already spoken to for this very book. "The only systematic sampling we have is from the Forest Inventory Analysis of forested lands. Otherwise, it's EDDMapS and whatever datasets individual entities might have (that the rest of us aren't privy to unless they happen to upload the data into EDDMapS). To add to the problem, EDDMapS reports often lack area covered by a species," wrote Loewenstein. Aside from specific localized projects or possibly some individual agencies/localities mapping what is on the acreage they manage, her guess is that most agencies rely on Forest Inventory Analysis data, EDDMapS, and the Alabama Invasive Plant Council. The way Loewenstein sees it, it's bizarre that so little insight exists on such a culturally iconic plant. "There really isn't very much research on seed fertility and fecundity, and how long the seeds last in the seed bank," she says in our interview preceding this exchange. "[Even] the basic stuff: how quickly does it spread?" What seems like a simple question lacks a clear answer. Reported estimates of expansion fall anywhere between 2,500 acres and 150,000 acres every year. Invasive plant specialists will uniformly tell you to check the data published by the US Forest Service—but that database doesn't include any details on the plant's rate of spread, only estimated total land coverage (which also notably has many gaps).

Loewenstein theorizes that obtaining funding is a challenge for researchers who are interested in studying kudzu, especially today, considering how other invasive plants like the rapidly spreading privet and Chinese tallow tree may be ranked as more of a threat to native ecosystems. "It's ironic, but we know about it," she tells me. "People assume that it's studied, and so people may be focusing more on the new players that are coming on the scene." After reaching out to the Alabama Department of Agriculture and Industry, Alabama State Parks, Alabama Forestry Commission, Alabama Department of Transportation, Alabama Farmers Federation, and the National Resources Conservation Service, I am redirected to four other representatives, three of whom are at agencies I have already contacted, until all inquiries encounter a dead end. The universal gist of the responses can best be encapsulated by three little, mighty, words: "I don't know."

The hunt for answers continues. Over in Texas, a public records request filed with the Texas Department of Agriculture leaves me yet again empty-handed, as the agency maintains it does not have data on kudzu. They instead send me a paragraph that quotes directly from a 2015 *Smithsonian* article that draws on incomplete FIA data to report kudzu's coverage and rate of spread: "Kudzu is typically said to cover seven million to nine million acres across the United States. But scientists reassessing kudzu's spread have found that it's nothing like that." At this point, I am redirected to the Texas Forestry Commission and the US Forest Service. Right back to where I began.

Louisiana is next, where the press secretary at the Louisiana Department of Agriculture and Forestry connects me with Hallie Dozier, an assistant professor of the forestry extension of natural resources at Louisiana State University and invasive species specialist. After she notes that there was an aquatic invasive species group that formed nearly twenty years ago and that state organizations decided to only focus on aquatic species or species involved with shipping, she clarifies that kudzu failed to make the list. Dozier tells me, in no uncertain terms, that if the Louisiana Department of Agriculture and Forestry cannot answer the question, she is not sure who can. Plot twist: They can't.

In Mississippi, I am redirected from the state Department of Agriculture and Commerce—which affirms that it doesn't track information on kudzu—to John Byrd, a weed scientist at Mississippi State University. Byrd tells me that the funding he's been able to secure in his research relating to the perennial has been focused on control strategies, not measurements of acreage. Byrd surmises it's impossible to know how much the state has spent on kudzu control, given the likely divide in investment of control between state agencies like the Mississippi Department of Transportation; Mississippi Wildlife, Fisheries and Parks; and Mississippi Forestry Commission, reaffirming what I already suspected. Everything is localized, dependent on specific places and their fragmented processes, and not one person or organization is aware of how much of the vine covers the state—nor do they know how much has been spent trying to curtail it.

Voyage unrelenting, I make a pit stop in Virginia, where I am connected with Kevin Heffernan, a stewardship biologist at the Virginia Department of Conservation and Recreation. Over email, Heffernan tells me that there is no single agency at any level of government that is solely responsible for tracking and recording the spread of invasive species. According to the biologist, there are different jurisdictions for different organisms based on their impacts on economic or natural resources—this has to do with the vast impacts of invasives. At its root, Heffernan says kudzu overlaps several jurisdictions, spanning impacts on agriculture, transportation rights-of-way, and public and private land. As for insight into how much land it covers, the state of Virginia also has—you guessed it—glaring gaps in what data they have collected. Heffernan confirms there is no precise information available on the cover of kudzu across the state, explaining their sources include the Digital Atlas of the Virginia Flora, which depicts the distribution of a species by county. The biologist affirms they also review reports of species using digital sites like EDDMapS and iNaturalist.org. Unsurprisingly, at this point there is no reported total for what Virginia state agencies spend on control of kudzu—which exists for other pests like the spotted lanternfly.[1] But he is not without hope that these questions are unanswerable—he reckons that remote imagery will one day soon be used to provide these missing precise measurements of kudzu cover statewide.

Next stop, North Carolina. According to Philip Jackson, a public information officer at the North Carolina Forest Service, the amount of land in the state covered by kudzu—and the amount of money invested by the state in controlling the spread of the plant—are currently unknown. Jackson also notes that not one entity is tracking and recording the growth of the species. Another stalemate.

With South Carolina just a few hours away, I head there next. After filing a FOIA request, I hear from Alden Dalton at the South Carolina Department

---

1. The Virginia Department of Agriculture and Consumer Services spotted lanternfly program oversees surveys and mapping of the invasive pest, as well as providing digital resources and training on how to manage populations.

of Agriculture. Dalton says the state is unique in the sense that the Department of Agriculture doesn't oversee the plant department; Clemson University does. At Clemson, I am connected with David Coyle, Clemson Extension Forest Health and Invasive Species specialist, who compares the abundance of kudzu in South Carolina to neighboring Georgia. He reports in an email that based on his anecdotal observations driving around both states, the amount of kudzu in Georgia is pretty similar, or perhaps even a little smaller in terms of proportion, to South Carolina. From there, I am rerouted to the South Carolina Forestry Commission for information on mitigation and removal efforts in the state, as well as economic outcomes, along with the South Carolina Department of Transportation, South Carolina Department of Natural Resources, and companies like CSX Corporation and Norfolk Southern (the latter of which Coyle anecdotally says privately managed to get rid of the abundance of kudzu that grows along railroad tracks across the state). At last I connect with David Jenkins, an entomologist and forest health program manager at the South Carolina Forestry Commission. Jenkins is unable to give me any statistics and does not know the costs of kudzu removal and mitigation in the state. He does clarify that the South Carolina Forestry Commission has not invested in controlling kudzu's spread, except to manage it when planting a forest.

Heading over to Tennessee also yields little traction. Samantha Jean, a director of communications at the state Department of Agriculture, confirms their agency does not keep records on kudzu and recommends I reach out to the Tennessee Department of the Environment—where I am met with radio silence, despite repeated attempts to get in touch with staff. Kentucky is another dead end—public record requests aren't available to people not living in the Commonwealth, and several dogged attempts at reaching state agencies are left resolutely unanswered. In despair, I move on to Arkansas, where I hear from Sarah Cato, a public information officer at the Arkansas Department of Agriculture. She shares a resource published in 2013 by the Arkansas Vascular Flora Project, the *Atlas of the Vascular Plants of Arkansas,* which shows that kudzu can be found in more than half of the state's seventy-five counties, with much of it concentrated in the Delta, an area of land flanking the Mississippi River, and along Crowley's Ridge, a geological formation made up of narrow hills that goes on for miles. Aside from that, the agency doesn't offer any new insight. From there, Cato connects me with Theo Witsell, an ecologist and chief of research at the Arkansas Natural Heritage Commission. He states that he is confident that no one has any idea how much land in Arkansas is covered with kudzu.

My odyssey ends in Florida, where the elusive vine spans an inconclusive number of acres. Listed by the Florida Exotic Pest Plant Council as a Category I invasive species, it was first introduced to Florida's Panhandle in the 1920s, since spreading southbound. Florida's Invasive Plant Management, operated out of the Florida Fish and Wildlife Conservation Commission (FWC), is the agency responsible for statewide programs controlling invasive species. But it is not tracking kudzu growth; nor has it ever done so. A public records request filed in the spring of 2022 confirms that there are no statewide public and private land surveys in Florida for kudzu. The state is the only one able to provide insight into the economic impact of management and eradication—the FWC funded $57,124.79 of kudzu control from 2017 to 2022. When asked about the state agency's information on kudzu land coverage and growth, FWC public information specialist Melissa Williams tells me in an email, on behalf of the state's invasive plant management team, that kudzu doesn't cause major problems in Florida like it does in other states, which Williams suggests could be because temperatures get too hot for its growth cycle. The same answer is given when directed to the USDA agency operating out of Florida, which tells me that while kudzu is an undeniably terrible problem for much of the South, the USDA-ARS Invasive Plant Research has no active projects for it. That's at least according to Melissa Smith, a research ecologist at the USDA's Agricultural Research Service Invasive Plant Research Laboratory. Smith clarifies that the lab, based in Fort Lauderdale, has never conducted any research on kudzu and that most of their projects focus on plant invasions with a fairly large footprint in Florida. But according to the University of South Florida's Institute for Systematic Botany, the plant's presence has been verified in more than half of the state: 39 of Florida's 67 counties, including Miami-Dade.

Together, all of these responses and reports are evidence that when it comes to kudzu, a patchwork system of accountability results in no one knowing much of anything. No single agency or source is responsible for tracking and recording its growth or economic impact as an invasive species. And perhaps in part because of that, the weed is no longer a priority for scientists and policymakers alike.

If you ask environmental scientist Paulina Harron, it should be the opposite—if for no other reason than climate change's expected impact on kudzu expansion and that formidable price tag. She is the lead author of a 2020 paper published in *PLOS One* on kudzu's economic impact in Oklahoma. Her research found that over the next five years, the spread of kudzu could result in a loss of $167.9 million and impact up to 780 jobs in the state's forest product industry. Harron's research revealed that kudzu is already costing the Oklahoma forest product industry $500 per hectare per year to control the infestation. On a national scale, the projected financial impacts balloon out. The most up-to-date cost estimates of US kudzu control—published in 2003 and 2004—land at around $200 per acre, every year for a five-year-period. A 1993 study by the Congressional Office of Technology Assessment reported that kudzu accounted for a loss of about $50 million annually across the agricultural and forestry industries. These numbers give just a glimpse at the scale of economic effect wrought by biological invasions or when an organism establishes a population in an area beyond its native range. A 2021 study published in *Science of the Total Environment* estimated the costs from 1960 to 2020 caused by invasive species to land somewhere around $1.22 trillion, with the agriculture industry being hit the hardest. More than two-thirds of the observed losses were in the form of damages, not management. In 2020, the US Department of the Interior earmarked $143 million for invasive species management—and from 1960 to 2020, the US spent $1.35 billion on preemptive measures like invasive biosecurity. After mammals, plants were the second most economically damaging—costing $190.45 billion in impacts, largely within agriculture. Alone, kudzu packs a pricey punch—the forestry industry spends an estimated $100 to $500 million per year on kudzu-related impacts. And an estimated $1.5 million per year has been spent by electric companies trying to manage kudzu growing on utility poles. "These economic impacts definitely serve as an incentive for governments, at different levels, to look into control strategies," Harron observed.

After venturing through numerous public records requests and navigating dozens of agencies across the eleven states that make up the South, I found myself probing a new question: Without data measuring the depth of its coverage, how can policy and research around the impacts of the vine, as well as classification determining the extent of the problem, be accurately informed? The answer is as twisted as a plot of eager weeds. According to invasive species specialist Stephen Enloe, it all comes back to what every other researcher in the space has already told me. "You will realize that kudzu is very poorly studied on that front. Observationally, one only needs to drive around the Southeast to see those big monotypic stands of kudzu where literally not much else is growing," said Enloe. "But in terms of the quantitative peer-reviewed scientific literature, there's not a ton of papers out there looking at kudzu's impact over time on biodiversity."

Old questions persist, dancing around my brain like a foraging honeybee. I can't seem to shake my growing bewilderment at the way this prolific weed has been written off with little evidence to back that up. I'm as rooted in this monumental missing insight as mature kudzu populations long settled into the abandoned fields of the South. I fire off a series of follow-ups: Who is responsible for maintaining these records, tracking kudzu's national and state-wide spread, as well as eradicating troublesome kudzu infestations across the country? Does *anyone* actually know how much of the South kudzu covers and what climate change means for its growth? "I don't think you're going to find a super accurate number that encompasses all non-crop land and historic disturbed areas where kudzu persists," Enloe replied. This is underscored as a problem not just relative to one meteoric vine. "It's an issue with invasive plant species in general, from a federal perspective of multi-state understanding," he continued. "Federally, there's not a really solid comprehensive approach that has been utilized. It's extremely difficult and expensive, surveying and mapping on relatively small areas, in a nation as big as the continental US. It's certainly a huge task." Enloe's answer to my next question came as no surprise. Is anyone actually accountable for tracking, reporting, and mitigating kudzu in the American South? "I'm not sure of any remote sensing efforts that have truly attempted to quantify total kudzu coverage across the Southeast at this time, and it seems like that would be the most logical approach to do right now. But I don't know of anybody who's actually done it," Enloe said. And no one has—at least not yet.

An emerging body of research suggests this may not be the case for much longer. A 2021 study in *Remote Sensing* tested a new classification model in Knox County, Tennessee, which combined linear unmixing, phenology-based mixing and nonlinear unmixing, to more accurately map kudzu with multi-spectral images at a county-wide scale. They found that the plant persisted in "small patches along forest edges, roads, and vegetation tops near houses and infrastructure" across the area analyzed, with the highest degrees of concentration in the northwestern and southeastern parts of Knox County. And a 2020 paper in *Data Fusion for Improved Forest Inventories and Planning* looked to map kudzu's distribution also in the geographical center of the Great Valley of East Tennessee by using machine learning algorithms and object-based classification with hyperspectral and high-resolution images. Observed limitations of that approach include "computationally intensive" algorithms, which makes them "inapplicable to large areas and data sets" and only viable for long-term monitoring efforts. Other noted barriers linked to mapping kudzu with remote sensing methods include accuracy issues in classifying the vine that stem from its similarity in spectral information to other vegetation, and the way kudzu's growth on top of a canopy makes it less spatially distinct from the plants it overtakes. Meanwhile, all of these efforts are limited to surveying smaller parcels of land.

Until this technology is up to the task, researchers will continue to rely on incomplete data. According to Enloe, the Forest Inventory Analysis Data is probably one of the best data sets out there when it comes to kudzu and other invasive species on forested lands. But he stresses that these figures can't capture the depth of the problem kudzu poses, as the surveys haven't captured where kudzu persists. "And so, we've often used that as a surrogate for true reporting," he clarified. Enloe also mentions citizen science approaches to reporting on invasive species abundance but cautions that because the very nature of these systems is voluntary, it won't be a "super accurate" thing.

Nearly all of the state agencies I contacted with requests for data on kudzu referenced a reliance on EDDMapS, which is operated out of the University of Georgia. Powered by reports of kudzu seen in the wild, and informed by federal land surveys of the vine's presence on forested land, the website can provide a snapshot of its growth and where it's been spotted.[2] That same resource keeps track of the states that have kudzu on their noxious weed lists or have legislation in effect banning the planting of kudzu. As of 2022, eleven states in

2. It's important to remember that this data is reflective of visual sightings of kudzu by citizen scientists and informed by federal surveys of forested land, which are not representative of the full scope of coverage. "Early Detection & Distribution Mapping System," University of Georgia, n.d., Center for Invasive Species and Ecosystem Health, https://www.eddmaps.org/.

the US have a law against cultivating kudzu, eight have included it on a state list of noxious weeds, and five have it on both.

Chuck Bargeron, director of the Center for Invasive Species and Ecosystem Health at the University of Georgia, runs the database. A self-described "numbers guy," Bargeron is one of the best people to ask about the challenges and drawbacks in invasive species reporting. He created EDDMapS because of it. "The biggest obstacle is that when you do what we do and pull data from basically any location that we can find, to just get a better picture of where these species have been found in North America, you realize that nobody does things the same way," Bargeron said. "You always have the who, what, when, where; but you don't always have the why or the how much," he explained. "The 'why' and the 'how much' are critical when somebody starts analyzing a big body of data." Because of that, Bargeron says a question like "How much kudzu is there in the United States?" is nearly impossible to answer—even after "all the time that we've been doing this."

A 2021 EDDMapS report shared with me courtesy of Bargeron shows they have 7,198 records in their database of kudzu in the United States, totaling 136,554 acres; 5,388 records belong to the Southeast, totaling 84,642 acres. As for county coverage, there are 1,137 counties across the nation with documented sightings, with 652 of those concentrated in the Southeast. Numbers like these paint a picture of the "vine that ate the South" that suggests it is inconsequential in the grand scheme of things. But they are also incomplete. And until the full scope of kudzu's reign across the edges of forests and farms and fields is more than a mere snapshot, the depth of its presence can never really be known.

When it comes to kudzu, Bargeron says that's unlikely to ever change because of the vine's persistent overabundance. "The unfortunate part is that the truth is there's limited examples of when we've been able to completely eradicate things," he said. While better data—on how much land it covers and its spread—doesn't exist for kudzu, it does exist for other invasive species. The reason for the discrepancy, he notes, is the pool of extremely limited funding, scarce investment that goes to tracking other bothersome invasive species that have a better shot at actually being eradicated. "It's not as simple as, 'We're never going to eradicate kudzu.' It's that there is *so* much kudzu, that it is hard to map it all." When asked, he confirms that no agency has been responsible for measuring and tracking the plant's growth. "I'm trying to think if there's a caveat, where somebody could say that was wrong. But I can't think of one." Part of the reason is that kudzu is no longer listed on the Federal Noxious Weed List. It was added in 1997 but removed in 2000. "I think kudzu was added for political reasons, and it was taken off because it didn't fit—it was too widespread to continue to be listed," he said. USDA prerequisites for classification on the list are clear; a species can only remain on the list if there's a fighting chance it can be eliminated. Without being on that list, Bargeron says there's little chance the federal agency will use any of its limited funding for a plant that doesn't make the cut.

When it comes to the Federal Noxious Weed List, there's no better resource than Anne LeBrun. National Policy Manager in the US Department of Agriculture's Animal and Plant Health Inspection Service (APHIS), LeBrun breaks down why a plant would be removed from the Federal Noxious Weed list in an email. She explains that the official definition of a noxious weed as any plant or plant product that can directly or indirectly injure or cause damage to crops—which she notes includes nursery stock or plant products—livestock, poultry, or other interests of agriculture, irrigation, navigation, the natural resources of the United States, the public health, or the environment. Removal can be petitioned by one person or entity, LeBrun clarifies. And some of the requirements for the removal of a plant from this classification include evidence that the species is distributed throughout its potential range or has spread too far to implement effective control, as well as evidence that control efforts have been unsuccessful and further efforts are unlikely to succeed, according to LeBrun. It's at this point when it also becomes strikingly clear that the APHIS never listed kudzu as a Federal Noxious Weed in its original designation. LeBrun explains that in 1997, Congress amended the definition of a noxious weed in the Federal Noxious Weed Act of 1974 to include kudzu, or Public Law 105-86. So in 2000, when Congress passed the Plant Protection Act, consolidating several older acts, including the Federal Noxious Weed Act of 1974, it redefined "noxious weed" and excluded the 1997 reference to kudzu. At the time, the historical references to kudzu on APHIS platforms were responsive to the language written into law.

The year was 1999. Just days before his impeachment trial and ensuing acquittal, President Clinton famously signed into law the very first executive order on invasive species. That piece of legislation was pivotal, for it established the formation of the National Invasive Species Council (NISC), also known as the Invasive Species Council (ISC), a cohort of federal agency leaders intended to prevent, remove, and manage all invasive species in the US. In 2016, the legislation was updated and reinforced by an executive order issued toward the very end of the Obama administration, which directed members of the NISC to consider climate change, human and environmental health, and technological advancements in their work. "That group has not met very often. I'm not even sure that I can tell you when actually they last met," Bargeron said of the council. Published in spring 2022, the latest *NISC Annual Work Plan for Fiscal Year 2022* included a priority activity of improving information management by using federal programs to advance the "collation, analysis, and distribution of information and data related to invasive species" and producing a briefing paper compiling the types of data relevant for invasive species management and how it's used for decision-making.

From 1999 to 2019, the Invasive Species Advisory Committee (ISAC) operated as an advisory unit working adjacent to the National Invasive Species Council, tasked with a mission to help inform and answer questions—beyond the scope of what federal agencies do—as well as guide federal policy decisions related to invasive species. Bargeron was a member for six years. "I was termed out of it when the funding was pulled," he said. The funding was cut for that during the Trump administration, which defunded and disbanded the committee in 2019. When I first spoke to Bargeron in 2021, the committee was still disbanded, but Bargeron expressed his hope it would return soon. "I've heard rumors since Biden was elected that it will be brought back," he expressed. "That will be a good thing, if for no other reason than to just show that there is interest and work being done on invasive species and draw attention to the effort." Those rumors were, in fact, rooted in truth—on September 30, 2021, President Biden reestablished the very same committee disbanded during the prior administration. And on January 24, 2022, the US Department of the Interior announced that the agency would appoint twenty new members to the newly restored ISAC. Two weeks later, the US Department of Agriculture released news of more than $70 million in federal funding going to 372 projects under the Plant Protection Act's Section 7721 program across forty-nine states as well as Guam and Puerto Rico. The allocated funding will help the agency and its partners detect and respond to invasive plant pests and diseases—further signifying the administration's interest in elevating invasive species research.

In late 2022, the names of thirteen voting members nominated to the committee were publicized by the US Department of the Interior, with the first meeting of the council since its unraveling in 2019 taking place in spring of the following year. Bargeron is among the select few who were appointed to ISAC, also serving as interim chair. "Invasive species are as big of an environmental issue as climate change," he noted, emphasizing how consequences of invasive species spread include loss of biodiversity, as well as loss of endangered and threatened species. Examples of this abound, lists as long as mature kudzu roots, lifelines embedded deep within the core of the earth. The emerald ash borer, a beetle native to East Asia, was initially spotted in 2002 in southeastern Michigan, before spreading to at least thirty-six states and the District of Columbia, killing tens of millions of ash trees. According to Bargeron, the long-term impact of that is "huge."

Although kudzu is well-known across the American South, the weed is not covered by mainstream media in a way that will propel federal or state policymakers to allocate investment for research and mitigation. The same is true for most invasive plants and animals—invasive research in the United States is notoriously underfunded when compared to other scientific disciplines. "This is a high-enough environmental issue that it is something that there's awareness of," Bargeron said. With more public awareness, the funding problem pervasive in the field of invasive species could be resolved: "The public wanting to see action is how funding is allocated." Despite the lack of resources, he has spent most of his career working to address glaring, open-ended questions in the field of invasive species. Most have to do with gaps in necessary information—a deficit of insight into how quickly invasives like kudzu spread, where they persist as warming temperatures effect plant growth, and what the plant's range really is. "Part of the reason we are doing what we're doing is because there was that gap," he said. The EDDMapS program is funded by federal and state agencies. "Our goal is to be able to someday answer that question: 'How much kudzu is there?'" Until then, those wanting to know more are stuck in a waiting zone—doomed to remain without clarity until the kudzu curtains draping over the margins of the South are deemed enough of a problem to prioritize once more. "There's no concerted effort across the Southeast to do anything about kudzu," Bargeron declared.

A group hoping to change that is quietly coming onto the scene. Established in the summer of 2022, the Southeast Regional Invasive Species and Climate Change Management Network (SE RISCC) is an emerging initiative led by Deah Lieurance, an extension scientist in the Agronomy Department at the University of Florida; Brett Scheffers, a conservation ecologist and assistant professor at the University of Florida; and Wesley Daniel, a fishery biologist at the USGS Wetland and Aquatic Research Center. According to Lieurance, the group aims to bring together climate and invasion science research and researchers in a collective effort to reduce the compounding effects of invasive species and climate change. It's a joint effort through the Center for Invasive Species and Ecosystem Health at the University of Georgia College of Agricultural and Environmental Sciences and Warnell School of Forestry and Natural Resources. Organization on behalf of invasive species research, especially in light of rising temperatures and carbon dioxide emissions, also somewhat exists at the national level. The National Association of Invasive Plant Councils—connected to the Center for Invasive Species and Ecosystem Health at the University of Georgia—which Bargeron describes as "more of an informal group" that serves as a forum for state groups to talk about what they're doing, is one source, while the North American Invasive Species Management Association (NAISMA) is a nonprofit formed in 1994. According to Bargeron, these two organizations in particular have ramped up their scope of impact over the last couple of years. He noted that they've "started to look at ways to influence and have a voice on policy in different ways." It's this type of lobbying work—on behalf of invasive species research—that he considers a "missing piece."

Of course, the biggest hurdle may come down to popular interest. "Invasive species have never been hot enough," said Bargeron. "I've joked many times: 'How many times is there a question about climate change in a presidential debate?' I just want to hear that question about Asian carp, or zebra mussels, or pythons, or kudzu brought up in a presidential debate."

11

CARBON CONUNDRUM

I have a confession to make. I am afraid of nature. It's a fear I'm ashamed of, but one that I understand. Unfettered wild spaces and places roaming with diverse species and flourishing, assorted plants are unfamiliar arenas for a girl forged in a sea of asphalt. Still, I wonder what my ancestors would think. The generations who came before me, carrying strands of my DNA, threaded into plot lines of their own. Those who celebrated the earth and all of its glory. I wonder what they would feel if they were to look down on the product of their greatness—here, in part, because of their lived sacrifices—spineless little me.

But nature is also afraid. It is fearful of the species that has waged war on its very being. And it has so many reasons to be. The natural world is under siege, humanity its greatest foe. Whether it's the climate crisis, careless industrialism, or a toxic alchemy of both, humans are the force behind every animal's and plant's race for survival. We are the reason for the calamitous loss of space and the dwindling resources hurting the planet. We have continued to threaten the existence of the very precious space hosting the billions of organisms that make up a single ecosystem. We have drifted so far from our innate connection and gratitude concerning the natural world. Instead of cherishing it and honoring it, as our ancestors once did, we've invested in destroying it.

Just as the aftermath of a hurricane varies wildly for a human population, the effects of the rapidly warming world on plant abundance, reproduction, and resilience look different for the hundreds of thousands of plants known to science. It can be hard to track the widespread impacts of this sort of crisis. One with a scope of the highest magnitude—overwhelming in the ways it reaches every area, every community, and every sector. It's even harder to measure what warming temperatures mean for something like kudzu, a plant that we know so little about. Rising global temperatures are triggering opportunities for new invasive species introduction pathways, range shifting, and faster plant growth. Droughts, storms, and floods are all climate extremes that amplify the likelihood of invasion expansion, while milder winters promote species survival and decrease the effectiveness of herbicides. Questions continue to abound, theories budding at the rate of a flowering path of kudzu: How *do* warming temperatures, compounding with heat waves, storms, flooding, and droughts, impact plant growth, resilience, and reproduction?

One notable American researcher is actively looking for answers. His voyage with kudzu can be traced back to the 1980s, when plant physiologist Lewis Ziska first crossed paths with the decadent, hulking presence haunting the southern countryside. He was in his late twenties, meeting his future in-laws for the first time outside of Tuscaloosa, Alabama. "As a plant person, you see trees and you see bushes and you see grass," said Ziska. It was there, deep in western Alabama, with his soon-to-be wife and her family, where he first noticed the plant that has entwined itself in the fabric of southern heritage. "But wait—what the hell is this? It's growing everywhere," he remembered thinking: "[It's] a solid green sheet of vine." Like a plant instinctively drawn to the strongest light, or a moth flirting with an addictive flame, he found himself instantly curious about kudzu and how it ended up in Alabama, emerging from every corner, swaths as tall as a tsunami wrapping the edges of forests in an almost intimate, striking embrace. "Other weeds don't have that same visual that kudzu has, and that's one of the reasons why we're so fascinated with it, its size, its propensity to grow, its ability to smother. It's incredible," he said. "You can't ignore it." From there, he dug into its origin and history, reading about the ways it's been applied in everything from basket weaving to medicine. "It's native to East Asia, but part of the southern heritage," he mused. "It's fascinating to see how in your face it is, in terms of its ability to disrupt natural systems, and how humans adapt to it, not only to recognize that it's a threat, but also to use it, at times, as a resource."

Culturally and ecologically, Ziska can't get enough of the weed that suppresses all other plants with its deadly shading. "Kudzu is one of those things I think would make a wonderful 1950s black-and-white science fiction movie. It's *The Day of the Triffids* science fiction thing, because it grows so quickly," said Ziska. A 1951 post-apocalyptic novel about an aggressive species of plant that devours people, John Wyndham's *The Day of the Triffids* was published during a time when public consensus on kudzu was changing—the federal government would blacklist the vine just two years later from its list of recommended solutions to soil erosion. "It would be the one where 'The Thing' comes in, the vine comes into your window and grabs your dog," he said. At sixty-four, Ziska has contributed to at least seven published research papers analyzing kudzu in the South—all within the context of climate change. An associate professor at Columbia University, he has authored more than 150 academic publications on rising temperatures and carbon-dioxide emissions impacts on everything from invasive species to agriculture. He's also written several books. *Greenhouse Planet,* his latest, takes a look at what rising carbon dioxide concentrations signify for plants, people, and ecosystems.

What we know about the impacts of climate change on something like kudzu is based on just a dozen experiments, according to Ziska. What is more extensively studied is the impact of rising carbon dioxide and climate change on selection factors for invasive species. A 2019 study he coauthored in *Invasive Plant Science and Management* reviewed existing literature to find that increasing CO_2 levels and invasive weeds' rapid adaptive evolution mean plants like kudzu are likely to come out on top. "It can select them over other species. And in the case of poison ivy, or in the case of kudzu, we think part of that is the morphology," he explained. "If I'm a tree, I've got to put some carbon aside to make support structures, right? I need the wood in order to get up high. But if I'm a vine, I don't need that. I'm going to take advantage of the trees. I'm going to wrap myself around them so I can get to the sun. So all the additional carbon dioxide, none of it is going for support structures. All of it is going for growth, new leads, new shoots." And a 2016 paper in *Biological Conservation* found that warming temperatures mean upwards of one hundred new invasive plants could be established in the northeastern US and that many invasive plants are also expected to expand known ranges under existing climate conditions. Intensifying heatwaves, storms, floods, and droughts fueled by climate-warming pollution are producing resounding effects for plant growth, resilience, and reproduction across all species. Like the dark, muddy waters in a Florida swamp, where kudzu fits into the picture remains opaque. The emerging research indicates that kudzu and the other vines it lives alongside are strong responders to rising carbon dioxide. "It's not a one-size-fits-all response," he cautioned, noting the roles that CO_2 and temperature play in plant biology. While other factors are vague, what is clear is that carbon dioxide will play a selective role, benefiting some species more than others. "So far, all of the evidence indicates that kudzu is going to be one of the winners in that selection process," said Ziska. The plant is poised to be one of the species that thrives in our imperiled world.

One of those experiments was led by plant ecophysiologist Heather Coiner, who found that an increase in winter temperatures shifted where kudzu was established. Ziska collaborated on that research. "We can't say with 100% accuracy that if I see kudzu in DeKalb County and Wisconsin, that everything between there and the Gulf of Mexico is covered with kudzu," he noted. Consisting of bustling metropolises like Atlanta, DeKalb County sits in the central northern part of Georgia—nine hundred miles south of "America's Dairyland." "It's not clear. But it is changing in terms of its demography, of where North, South, East and West, it's starting to show up. And we think that there is a role that climate may play in that." Ziska is hesitant to speak conclusively about these trends—largely due to the limited research that's been conducted to explore these patterns further. Questions continue to flourish about the distinctive plant, especially in terms of its rate of spread and capacity for growth. "That being said, the actual number of acres that are covered is a best guess, basically. And it's true for cheatgrass, it's true for many invasives. It's hard to get a handle on it," he said. "This is one of the things that we need to have the resources to do, is to not only just look at the damage that this is doing, but also get a sense of how quickly it is spreading and what's being vulnerable, what's being under threat. And we just don't have that yet."

To Ziska, finding answers to these questions for invasive plants like kudzu is imperative—especially in our warming world. There are two arguments for it; the first is purely ecological. "If you want to do the tree hugger argument, you can say, 'Well, yes, obviously, when kudzu comes in, it has a hell of an effect in terms of biodiversity,'" said Ziska. "There might be one hundred species in your forest, but once kudzu arises, there's only one. Kudzu grows over everything, shapes everything. Whatever animals [and] insects rely on those one hundred plant species, they're gone." There's also the economic rationalization for learning more about the vine and the way it interacts with the world around it. "What happens to my power lines when kudzu is growing on them? What happens to my harvestable forest when kudzu is growing on it? That costs me money. It costs me money to control," he added. According to the researcher, there are strong environmental, ecological, and economic arguments for wanting to control and understand how climate is going to affect invasive species biology.

The lack of funding interest is a universal foe faced by Ziska's peers, as well as emerging researchers wanting to scrutinize the complexities of a vine with a storied past. As is often the case with invasive research, a dearth of resources and the sheer scope of a project that aims to assess what warming temperatures mean for a plant are impassable hurdles in the investigation of kudzu. "That's part of our shortcomings. We don't have the funds. We don't have the resources to set up these real-time experiments," expressed Ziska. "The best we can do is to look at it from a historical perspective, sometimes even from an herbarium perspective." One typical strategy is planting seeds where they collect different bits and pieces of plants over a period to see how they change over time. "It's been very frustrating, but at the same time, very fascinating," he asserted. Invasive plant research is still a budding field. Even in its infancy, common emerging themes are identified by specialists like Ziska and by everyday people who have devoted their careers to analyzing the nuances of plants—what they offer and what they take away. If you ask the scientist, he believes the lack of concern about the spread of vines like kudzu comes back to human instinct—or a lack thereof. "As humans, we have the attention span of a hummingbird on crack. If we don't see something happening right away, we tend to ignore it. And so often for invasives that's what happens," he said. "We don't think about it until it gets to the point where 'Oh, what? Where did all this bamboo come from? Oh wait, why do we have greening disease in citrus in Florida? Or what's happening with soybean rust?' All of these things happen almost overnight."

But they happen in such a way—in a cycle that appears doomed to keep repeating—that Ziska suspects we're not psychologically tuned to. "We pick up on animals and things that move those invasives much more quickly and we're concerned about it. I bet you will find better numbers for the Burmese pythons in Florida than you will track for kudzu because that's how our minds work," he said. "No one drives a car down [Interstate] 95[1] and looks up and says, 'Oh look at that new cedar out there. I've never seen that before.' We see deer, we see things that move because that's what's in our wheelhouse. That's what we're programmed to see. We don't see plants, we don't see diseases, we don't see insects in the same way." In terms of responding to the threats that invasives pose, he thinks that's a byproduct of the obvious disconnect: "The ignorance, if you will, of human beings. We don't understand the plant community."

It's that animal versus plant dichotomy he keeps going back to. To Ziska, the typical societal response to vegetation is puzzling—especially given its prevalence in human life and the habitats we rely on. "If you weighed all the biomass on Earth, and divided it up into animals and plants, well, plants are 90% of what is existing out there. They have incredible genetic diversity, incredible chemistry, incredible life force," he noted. Plants can even contain more DNA than humans, with some species of ferns carrying more than twenty times the amount of chromosomes that people have. "But because we're animals, we have a different mindset of how we respond to plants," he added. That disillusioned response is where things get murky—and is exactly why, after centuries of being active in the southern landscape, there's very little insight into the inner workings of one formidable, instantly recognizable vine. "At the same time, it's one that we perpetuate, it's one that we want to grow and it's not the first of its kind. We have other plants, other insects, other things that we brought in on purpose to try and shift the whole biogeography of the world around [us]. And sometimes we pay a very heavy price for that."

Rather than continue to ignore the presence and power of kudzu, Ziska hopes we as a society can learn to understand it, to use it, to appreciate it— and even to respect it. "That's what we recognize as human systems. We don't see plants per se, but when that's the only plant there and it's nothing but this huge, wonderful vine that's growing over anything and everything, it's hard to ignore," he said. Perhaps the weed could soon serve as a vessel of meaning— a forewarning that encourages people to pay more attention to what climate

1. I-95 is an East Coast highway that stretches from Florida through Maine to Canada.

change means for invasive plants and why this relationship matters. "If you see something growing over a barn, or a car or a building, it's really hard to ignore it. You see it as an invasive. You see how powerful it is," the scientist mused. "It's a good prototype for people, who may not know anything about invasives but have seen kudzu, to understand that link to climate change." When it comes to developing that narrative around an invasive that people will recognize, Ziska thinks the scapegoated vine has a better chance than most of its leafy, weedy cousins to make a significant cultural impact. "It makes a good story," he said.

But how exactly the rapidly rising temperatures and carbon dioxide emissions will influence kudzu growth and range expansion is something no one yet has an explicit answer for. Even the federal agency in charge of invasive species management and monitoring across the US is at a loss. "We just don't know," admitted the USDA research ecologist Qinfeng Guo. "In the future, if climate change continues, warming continues, then what happens? Kudzu could be more abundant," he said. The handful of studies that exist support that thought. One can just look at the expansion of its range—what was once solely a southern problem is now spreading elsewhere, as kudzu has recently branched out northward into Pennsylvania and New York, been documented as far north as Maine and Washington, and spotted crossing international borders into Ontario and Nova Scotia. Populations have even been reported in Texas, Oklahoma, Iowa, and Nebraska. *Why* this is the case is where things get a little complicated. What scientists do know is that ranges of kudzu have naturally waxed and waned over the years, influenced by the weather—but those conditions look starkly different now. Our rapidly warming world has resulted in a crucial reduction in crop- and tree-killing freeze events, with climate change causing a poleward shift in the range of a multitude of invasive species. For the native flora that succumb to fellow plants like kudzu, what a human-induced changing climate means for native ecology may not just be a temporary concept, but a permanent foreboding. Still, some researchers are split on the poleward movement of the vine, as the handful of peer-reviewed studies done on the projected expansion have produced conflicting results.

Those are also sporadic. The people investigating the vine's connection with climate change can be counted on one hand. Plant ecophysiologist Nishanth Tharayil is among that group. Thanks to his work, we know that kudzu can impact not just the soil it invades but the nitrogen cycle of air around it. In a 2014 study in *New Phytologist,* Tharayil and Mioko Tamura found that kudzu results in the release of 4.8 tonnes of carbon annually. (*Grist's* Jim Meyer reported this is equal to the amount of carbon stored in 11.8 million acres of forest in the US.) Examining the impact of invasion on soil carbon storage in native ecosystems through the lens of one of America's most widespread invasive plants, the researchers looked at how increasing temperatures are impacting a plant's capacity for carbon sequestration—or in the case of kudzu, carbon release. They purposefully selected two very conflicting and contrasting invasive species for this study: kudzu and Japanese knotweed. Japanese knotweed is the antithesis of kudzu in the way it sequesters carbon. It releases CO_2 back into the atmosphere, while the "vine that ate the South" does the opposite. Rising temperatures are accelerating plant litter decomposition, which directly impacts underground carbon storage; the faster the decomposition, the more carbon is released that's stored in the soil. Of the two, they found that as the climate continues to warm, rising temperatures will have more of an impact on Japanese knotweed. With kudzu populations, the impact is less significant because there's already a rapid rate of decomposition.

The effect of this goes beyond invasive species; it's about the influence that plants exert on soil carbon storage. Kudzu and knotweed could be useful in the long run in the role they could play in capturing carbon as the world rapidly warms. Each weed facilitates the decomposition of grasses, which makes the product of the decomposition that is the side organic matter much tougher—meaning that they can both withstand an increase in temperature and aren't prone to oxidation. Published nearly a decade ago, this research was intended as a way to contrast the cyclical impacts of two invasive plants on carbon, investigating how the chemistry of what these species produce would alter soil carbon storage and how that carbon stored in the soil would interact with rising temperatures. In an agricultural system where we want to promote long-term carbon sequestration, if we were to just ignore the fact that kudzu and knotweed are invasive and let them grow as an operation of leguminous crops, their growth could fix more atmospheric nitrogen. Like little carbon sinks, these plants could help reduce the amount of carbon in the air around us.

But the imposing weed doesn't just pack away carbon—it also contributes to ozone pollution. Four years before that paper was published, a 2010 study in *Proceedings of the National Academy of Sciences* found that kudzu in the American South may be raising ozone levels in the region. An adjunct associate research scientist at the Earth Institute at Columbia University, Jonathan Hickman was the lead author in what was one of the first pieces of evidence to link kudzu with worsening air quality. In his research, he demonstrated that the vine also affects atmospheric chemistry by contributing to surface ozone pollution, which is both a potential health threat to humans and a deterrent to the growth of other plants and crops. Raised in Maryland, Hickman can't recall his introduction to the bewitching vine. One day, it was just there, something he was aware of, knew little about, but still instinctively regarded with fear. "It had a mythic quality to it," said Hickman. "There are all these stories about it coming in through a window, and strangling a baby in the crib, and all that stuff," he said. To date, the fifty-year-old has published two peer-reviewed studies on kudzu's impact on nitrogen oxides, or NOx, in the atmosphere, in the interest of testing how kudzu was affecting ecosystems. The body of research on kudzu from an ecosystem and nutrient-cycling perspective was almost nonexistent when he first set out to investigate it; he was able to find some work on how kudzu changes microbial communities, nitrogen cycling in the soils, and how it was changing the hydrological cycle, but nothing beyond that. "No one had really looked at how it was affecting the atmosphere or atmospheric composition," he noted, speaking to the lack of quantitative data, which was applicable to all invasive plants at the time of his research. That missing gap of insight on a vine with a complex cultural past was all the researcher needed to start digging for answers. "The fact that this was this wild and crazy poster child for invasive species," he said, remembering his reaction to the lack of evidence. "To have this invasive plant that is so well-known, and for no one else to have looked at that. I kept thinking I could find the paper, because it was such a clear and obvious question. And then, come to find out, nobody had."

To start, Hickman headed down to Georgia, a hotbed for the "scourge of the South." While there, a friend let him use their University of Georgia lab as a home base. Once he was set up, Hickman hit the road again. For a few days, the scientist drove around the state, seeking plots of flat land overwhelmed with colossal sheets of kudzu. He was trying to find sites that weren't too close to busy roads because the tailpipe emissions from combustion-engine cars, trucks, trains, and buses serve as a very large source of NOx, which would skew the results of his testing. Although he had no trouble locating sites overrun with the plant, it was a challenge locating a spot that could serve as his "control" plot—one that was kudzu-free but adjacent to those kudzu-enveloped zones. At long last, he narrowed down the elements he needed for his experiment: several sites spread out over three distinct locations in Georgia, about fifteen-minute drives from one another. Each had soils that were covered by kudzu and soils that were free of it. The results upheld his theory: Kudzu was indeed making a negative impact on the atmosphere, even intensifying the ozone pollution in the areas he observed. His research found that the presence of kudzu in a rural area, especially in an area without much fertilizer in use, likely contributes to increases in ozone. "Whether it's really bad for air quality is still an open question, but it's making a contribution, and it's one of those things where every little bit is adding up," said Hickman. What climate change's impact on this will be isn't clear yet. "It's probably making ozone pollution worse than it would if it wasn't there," he noted. How much worse? He doesn't know.

Hickman didn't stop there with his detective work on the impacts of a weed many have learned to love to hate. Over the course of three years, Hickman conducted fieldwork on kudzu. At one point, he even attempted to replicate the research he'd done in Georgia—following the thought process that it, along with countless other invasive plants, will move poleward as global temperatures continue to intensify in Maryland. Using a database provided by the Nature Conservancy, Hickman drove around to identify sites that he could use for his experiment. But his team didn't find the same effects on ecosystems that they had in the Deep South. They saw very limited effects on nitrogen cycling. The kudzu leaf litter had much higher nitrogen content than other species in the area, and it decomposed much more quickly. There was also no significant effect on greenhouse gas emissions. All of these results, according to Hickman, point to the need for more rigorous study of the vine.

Sandra Hoffberg, a National Science Foundation postdoctoral fellow at the Eaton Lab at Columbia University, would agree. The lead author of a 2016 study in the *Journal of the Torrey Botanical Society* that examined the poleward expansion of kudzu in the US, she spent twelve years in Georgia, where she pursued her PhD studies and wrote a thesis on kudzu—and got to know the weed that drapes across southern landscapes intimately. "I was collecting leaf tissue from kudzu and wisteria and people would pass by and say, 'What are you doing?' And I would say, 'Oh, I'm doing this,' and they were like, 'Oh, kudzu, that's the *worst* species. Oh, wisteria! That's my grandmother's favorite flower!'" Hoffberg quickly picked up on the differences in popular reception of the two, the sharp contrast in tone and the disproportionately negative responses leveled at kudzu. "People have these really strong opinions about these plants," she said. "These were plants that everybody wanted to give their story about."

The then-aspiring scientist spent much of her time in abandoned fields, parking lots, or forest edges across identified kudzu sites in New York City for her and coauthor Rodney Mauricio's study. Lost within spaces teeming with kudzu, she tracked down and visited twenty-two recorded sites dating back more than two decades to see if the planet still persists in those same spots today. Hoffberg wanted to challenge one of the assumptions about kudzu, that the vine never leaves an area it invades. "Everyone in the South was like, 'Yeah, of course it's still going to be there. It's always there, it doesn't move.' People's intuition about that is what we found in the core of the range, but it's not what we found in the North," she said. A born-and-raised New Yorker who went to school in the South, she points out that kudzu is not really as prevalent on the North's radar. Using data collected for a 1989 study published in the *Bulletin of the Torrey Botanical Club* that recorded the distribution of kudzu in New York City, Hoffberg discovered that 32% of the populations once recorded in the area that they could unambiguously identify still persisted. These results surprised her due to their longevity. She found there are some established, sizable populations of kudzu that have persisted for at least twenty-seven years in the surrounding area. But not every population was the same.

As the climate warms, the relentless vine is likely to continue to pop up in new northern ecosystems. "When it comes to climate change, it's not really a question of whether it's happening or not," declared Hoffberg, which is a takeaway supported by a greater than 99% consensus across all peer-reviewed scientific literature. According to NASA, the "vast majority of actively publishing climate scientists agree that humans are causing global warming and climate change." One of the latest Intergovernmental Panel on Climate Change reports—deemed "the most authoritative reports on the topic"—published in 2022 and prepared by 270 scientists across 67 countries, found that if global greenhouse gas emissions are not aggressively curbed, human-induced climate change is on track to wreak havoc on every sector of life. From public health to the economy to international security, everything is at risk. "The poleward shifts of species? It's not really a question of whether that's happening—it's very common," noted Hoffberg.

Climate modeling produced as early as the 1980s suggests that the New York metropolitan area should have been a suitable habitat for kudzu to grow and flourish. But she didn't see that in her study. "There are anecdotal references to kudzu being found even farther north than New York," she observed, adding that it doesn't seem like the weed is a huge problem in the state because the populations are not persisting over the long haul. Some limitations to this type of survey lie with whether any of the sites that no longer host kudzu have been populations eradicated by humans—in one example, a 1989 plot of kudzu land identified by Frankel turned out to be the home of a grocery store parking lot in 2016. "Is it because the landscape has changed such that the kudzu could not live here?" questioned Hoffberg.

Another impediment could be the location accuracy of the historically documented sites themselves. Phase one of the investigation focused on figuring out where exactly Frankel's described sites are in modern-day New York—using clues from the text, sans the help of any GPS coordinates. It took a little bit of studying Google Maps to try to find where exactly these sites that were published were, according to the scientist. From there, she incorporated census data into determining if the places where the vine was still prevalent—as well as noticeably absent—corresponded with the urban or rural settings for each location, compounding with population densities relative to each place. Although she didn't find that the vine greatly persisted in the northern range of its American presence, she found the opposite for sites farther south. Cross-comparing the New York dataset with herbarium records identifying nineteen sites in Georgia and South Carolina, when the plant was found to occur in the same approximate time period, led to Hoffberg finding that 95% of those original southern populations had persisted. "In the core of its range, kudzu is everywhere," she noted. Based on the research, the following conclusion can be drawn: In a rapidly changing climate, kudzu has kept right on flourishing in the American South but is far less likely to survive at the expanding edges of its own far-reaching borders. Hoffberg's final observations land us in a now-familiar spot: The tangled case of kudzu requires further probing. "This question about range expansion due to climate change? It's a simple question to pose, but it's not a simple question to answer," she said. "It is not always something that can be done in one study."

Momentum has been stymied across the invasive research community, however—and some, like Hoffberg, credit the lack of interest in further analyzing the weed to one prolific news article that changed the court of public opinion. In 2015, horticulturist and journalist Bill Finch published "The True Story of Kudzu, the Vine that Never Truly Ate the South," in *Smithsonian Magazine*. According to Hoffberg, the article presents kudzu as "not really a problem, and that everybody overreacted to kudzu." The way she sees it, that depiction is false, a social repackaging of a vine with an indisputable legacy of enthralling the American public. Per the story, kudzu's impact is sensationalized because of the very spaces it invades—old fields, abandoned lots, and edges of forests. In the piece where the unjust cultural demonization of kudzu has been—perhaps most famously—articulated in the US, Finch makes the case that the vine is less impactful than most have believed it to be for more than a century. He relies on the US Forest Service's Forest Inventory Analysis Unit's 2015 forested land estimate (which estimates there are 225,000 acres of kudzu across the South) to build his theory, concluding that kudzu does not present as significant an ecological problem, nor has it been as calamitous, as people have portrayed. In the years since its publication, the article has been widely circulated. On Google, it has the second-highest search engine optimization ranking when typing in the search term "kudzu"—out of more than six million results. Anecdotally, nearly every ecologist and invasive plant specialist I interview about kudzu brings it up. A few have nothing but praise for the piece. Others, like Hoffberg, imply it spreads misinformation. One of her critiques is that the author doesn't account for how the federal measurement of land coverage *doesn't* include the nonforested areas where kudzu most heavily persists. By that logic, Hoffberg questions whether the existing data on kudzu—and the figures the piece is framed around—can provide an accurate overview of the plant's true scope and spread. "The problem of the lack of quantitative studies on kudzu existed before this article, of course that article is not that old, but there may be this idea that if it's not in the forest, it's not that important, which I disagree with," she said.

The overwhelming response to the story is what really highlighted the lack of information and the lack of quantitative peer-reviewed research on kudzu to Hoffberg. She suspects that because of the piece's popularity—and the idea it touted that because kudzu invades disturbed habitats over pristine forests it's less of a dilemma—other researchers almost universally adopted that view and were dissuaded from looking further into the vine's propensity for growth and its ecological impacts. Another element that isn't addressed is any reference to climate change—or what rising temperatures and carbon dioxide emissions mean for the spread of the vine. The bulk of the body of research on kudzu shares a conclusion that where the species is invasive, damage is inflicted on the natural world around it. Still, thoroughly disseminated texts that challenge the idea of kudzu's widespread influence and minimize the weed's malignant qualities—pointing fingers at mainstream misperceptions of the plant as the only nuisance—have planted seeds of doubt. "The problem with this article in my eyes is we've already known that kudzu isn't in sunny areas," Hoffberg said. "It's everywhere except for the forest. But that's still a lot of space, and it's still important space."

Hickman, Tharayil, and Hoffberg have something in common: They all have a lot of unresolved questions about kudzu, which they collectively say should be further investigated. None are actively pursuing them. For Hickman, his experience studying the vine was complicated by a lack of resources. For Tharayil, the vine was never the focus of his research and likely never will be. For Hoffberg, kudzu served as a topic for her dissertation and, unlike her counterparts, she hopes to one day return to examining it. No more than a handful of researchers across the country are actively looking into learning more about the vine that once captured the hearts—and the pockets—of a rapt southern audience. And the one man who devoted much of his life to seeking answers is long out of the game.

Affectionately nicknamed "Dr. Kudzu," James Miller studied the vine for four decades. He has had a lot of nicknames, but early in his career, while looking at the plant many delight in denouncing, one of his colleagues flippantly labeled him "Dr. Kudzu." And just like that, the particular moniker became his calling card. "That was one of those nicknames that just became my name," Miller said. Ever since, he has leaned into the association. "Dr. Kudzu" is reflected in his published work, cemented in several news articles, and even remains in his email correspondence years after the scientist formally hung up his research mantle.

Born in Oklahoma, Miller and his wife Anne relocated from Indiana to Oregon before heading overseas to Southeast Asia, where they spent three years before moving to Alabama. It was there in 1977, at the US Forest Service station operating out of Auburn University, that he joined the federal service. Alabama would become a permanent home. These days, he lives in a forested expanse outside of Auburn. Like his peers, Miller has long sought after the elusive figure revealing how much of the United States kudzu impacts. Guesses abound, but as it stands, no one knows exactly how much land kudzu covers. What is known about the striking vine is that for little more than a century, one plant in a space made up of thousands has ruled as queen of the southern court, elevated in its ranks of weedlike infamy. "Dr. Kudzu" himself can't offer any new clarity into the puzzling, fraught nature of our entanglement with kudzu. When it comes to pressing questions in the body of research on the plant, like how much land it actually covers and how quickly it has spread, the retired invasive species specialist who spent decades working with it doesn't have answers. "Nobody has ever had that number," he said of kudzu's total land coverage. "I've given these interviews for forty years, I guess. We don't have that number."

It's clear he doesn't see the point in pursuing research in kudzu, despite the unsolved inquiries or the doubt and uncertainty that have remained embedded within our fractured understanding of how one plant works with the world around us. "What about the other invasive plants?" questioned Miller. In his view, even without the quantitative measurement to confirm how much land kudzu covers, kudzu is no longer a priority in the world of invasive plants. The weed is only a headache for whoever is directly dealing with it. "It's not a problem—it's a big world—unless it's on your land," he declared. It's a stance he's taken before. In a 2016 *Atlanta Journal-Constitution* article penned by journalist Dan Chapman, Miller was quoted saying that "it's in retreat" from the southern landscape it's long persisted in. Five years later, he backtracks on that opinion. Instead, he feels that "the kudzu story has taken another turn since that interview." "Well, it's not decreasing," Miller clarified. "Again, I could be dead wrong, or I could be dead right. Nobody knows." He does confirm, however, that he was wrong to tell the reporter he spoke with his estimate of how many acres remain in the South—500,000—but is otherwise unwilling to go into detail about *why* he gave that figure to begin with. "There's not a picture of it. There's not numbers. . . . There's just not. And so, anybody gives you a guesstimate? That's what it is," he said.

Decades ago, when the weed had a bold grip on southern pop culture, the scientist would receive correspondence from people across the region, everyone from farmers to widowed wives whose husbands used to eradicate it on their property. They were all reaching out to "Dr. Kudzu." "There's many of those letters," Miller recalled, noting that they continued to come throughout the years. "That was really heartbreaking. And that was a general southern folklore affair." In many ways, the vine is still stark and omnipresent in the ethos of the South. A little more subtle now, it manifests in language, in literature, in artistic displays, and within embedded cultural associations—as tough to eradicate as an unyielding swarm of kudzu.

At seventy-seven, Miller seems to want to leave his time investigating the consequential weed in the past. There's a hint of resignation in his tone when he describes the modern veil of mystique shrouding a plant that he orbited for all those years. It's a legacy that still follows him, as one of the American scientists behind the most peer-reviewed journal articles that look to answer questions about one unrelenting vine. Widely considered an expert on the perennial, Miller has dozens of academic publications to his name. According to ResearchGate, that body of work has garnered upwards of eight thousand reads and more than two thousand citations. And much of that delved into how to kill kudzu. "I'll tell you, this was really—this is a hard-pull interview. . . . This is not the thing I need to be doing," he declared. "Only, I will not give it up." Before retiring, Miller made appearances everywhere from gardening committees to research conferences in the last of his series of presentations on the "vine that ate the South." At each of these events, Miller remembers repeating the same universal message: "I would say, 'These are our new plants. These are our new plants, and they're not going to go away. So let's deal with it in that manner. And again, it's your problem, and you got to face up to it,'" he noted. "It is what it is. It's here to stay."

Although he's no longer particularly concerned about kudzu's dogged presence in the southern landscape, what Miller finds worrying is the role invasives play in a larger dystopian narrative—one where humans are at blame, the only defendants guilty of the ecosystem upheavals and environmental disturbances we have faced, are facing, and will continue to face. "Humans think they're rulers of the land," he said. "Every invasive . . . they've all been planted by humans and spread by humans, just like kudzu, but [we're] still doing it. Nobody's planting kudzu now, but they sure are planting silverthorn and Bradford pear." The former is a flowering shrub found in the southeastern US while the latter is an invasive deciduous tree congregated in both the eastern US and parts of the West. "You name it, they're planting it," pronounced Miller. The cyclic nature of it all is what aggravates the retired researcher—a feeling he's stuck with while witnessing the ways humanity keeps repeating the same mistakes over again. "We're still shooting ourselves in the foot, as I've always said," he added. In every sense, our causal relationship with species invasions also connects to our disregard for the health and preservation of the planet we live on. "Now what we're facing with the loss of species, it's just incredible, in every kind of life form. Plus the habitats, the ignorance of people who support them, how they live, connections to the land, connection to the whole atmosphere," argued Miller. "I mean, we just keep pumping it like it's a garbage disposal. This whole ignorance, it just aggravates me. We're in that age, and invasive plants are a part of that."

"Dr. Kudzu" himself has become a component of the century-old narrative attached to one of those very plants. Connected through the years of work he's done in the field of invasive species research, Miller says he isn't bonded to only one species, but all of them. Living in a tiny space nestled within a parcel of forested land near Auburn, on the outskirts of central Alabama, he spends his time among the plants and wildlife he's long revered, feeding something like a dozen wild deer a day, alongside other hungry neighbors, like raccoons. "People feel so stupidly bound to their landscapes and to having bigger carbon footprints. We have a very small home. We have a very small area in the forest and that's where we reside, and we don't want any more. We've taken too much anyway," he expressed. "I keep telling my wife when we live with spiders or whatever it is, this is their habitat. We're their guests. We're temporary guests."

Amid the flora and fauna that thrive in one nook of this Alabama forest, the Millers have carved out an existence where they live in tandem with both native and invasive species. Unlike the rest of modern civilization, they don't conquer the space as their own, but honor every element of the ecosystem they occupy. "I have an emotional attachment to it all. The earth. The sky. I've studied it all my life," he said.

"And it hurts me whenever somebody hurts this, out here. My home."

12

CONDEMNED

There's a point between the Great Smoky Mountains, the Nantahala National Forest, and the Pisgah National Forest in North Carolina that Casey Lance Brown calls home. The forty-three-year-old lives on a fourth-generation farm in a spot encircled by wild flora and fauna, which serve as both staples of a familiar environment and sources of inspiration. "It's all mountains, the whole thing. And there are valleys that run right through the middle. That's where the majority live," he said. His family has been living in Southern Appalachia for at least eight generations, according to Brown. It's the edge of civilization—a fitting domain for an artist whose muse is a plant that thrives on the outskirts. "All of us, we just live on the edges of those national, federal landscapes," he continued. "Plants have always been a big deal to me, because they surround me. They're like friends. They're my environment entirely."

A self-described "landscape futurist," Brown has been working in design research for about a decade. Landscape architecture is a common trade, but Brown says he is the first "futurist" he knows of. So much so, in fact, he coined that term himself. "Landscape architects are largely asked to work on projects that have value. Somebody is invested in you, as a designer, and you are hired to work on it. But I'm often looking for value that people can't see yet," he said. Like abandoned mines in the countryside, ramshackle buildings left in disarray, or industrial zones in long-deserted cities. Finding worth in undervalued places is his specialty. "It's like time travel," declared Brown. "The abandonment—you can see the old use, so that's the past—but it's still present. In the present, it's turning over to some sort of new use, which could include feral park space, right? Or wild plants and animals growing. And then that's its future. And you can see all those in one place at one time."

Brown was raised in a world of agriculture, one that introduced him early on to all kinds of vines, including kudzu. "My family, of course, were farmers," he said. "And so we dealt with managing plants my whole life, including dealing with weeds that grow on the edges of farms." He can't remember his first rendezvous with kudzu—it was always just there. Later in life, Brown would work for North Carolina's Mountain Research Station—operated by the state's Department of Agriculture and Consumer Services' Research Division—in Waynesville, North Carolina. Weeding and hacking at swaths of it, while managing some of those landscapes in between the USDA's Natural Resources Conservation Service and the Agricultural Research Service offices, is where he remembers first learning more about the "vine that swallowed the South."

"For me, it was always just 'these are plants that grow in annoying places,'" he recalled. Brown wrote a paper in graduate school about multiflora rose, which is another "one of those plants" that he cites as intentionally, and later lamentably, spread by federal agencies. While digging into multiflora rose, the lifelong resident of the South came across literature chronicling how kudzu was the "touchstone reference for environmental damage." "I remember definitely thinking, 'Well, that's questionable,'" noted Brown. Every summer while away at college, he would come back to his hometown in Appalachia. Studying landscape architecture and pursuing different artistic media, he would set off on photography excursions, exploring the terrain around him as he captured images of damaged and derelict sites dotted along the mountainous landscape. "I found this barn that's covered by kudzu, [and] I photographed it. I was like, 'Oh, this is awesome.' And I just kept thinking about it," he recalled. "I said, 'Man, it's like a monster.'" That's what led to the birth of "Kudzilla"—a visualization series where the artist uses a composite of images to appear like a real photograph of kudzu overtaking parts of the landscape. One of his Kudzilla images is on display in Raleigh—it's a haunting shot depicting a gas station being overtaken by kudzu. Such a place doesn't actually exist; in fact, it's nothing more than a product of Brown's creativity, the result of manipulated images of several gas stations fused with large-format imagery of a hillside consumed by kudzu close to where he lives. "It's artificial life—I've made it up," he said. "But I'm hopefully tricking you to some extent into thinking it is a photograph."

Born and bred in the South, Brown firmly believes kudzu has never been the crisis it's been stigmatized to be. If you ask him, all you need is a goat or two to keep an unruly patch of it under control. According to Brown, the weed has been selected as a scapegoat for a deeper problem; one where the reported overgrowth of a vine is used to fuel negative stereotypes and perceptions of his culture and his home. "It just seems that people often, especially the people that don't know kudzu natively, seem to use it to disparage regions where it grows," said Brown. He's talking about places in the North where people are only familiar with kudzu in southern settings. Brown expresses frustration with tourists lining up to see the "worst thing the South's ever done." "It represents uncivilized backwardness, right? You get these tourists, they're coming from the North: 'I'm gonna see the real wild South,' and they go traveling around, they see it all over the road. They're like, 'What is that monster?' They're like, 'That's kudzu. That's the worst thing anybody's ever done.' But they don't come back in the winter and see that it dies back," he added.

Centered on this concept of landscape futurism, Brown's visualization series on the "scourge of the South" relays a pointed message about the ways one invasive plant has been miscast in popular culture. "When we say the word kudzu, I don't think we're talking about the East Asian leguminous vine at all. I think we're talking about a phenomenon," he proposed. In his eyes, kudzu is a hyper-object—a philosophical concept that is too large to conceptualize, but not so abstract that you don't know what it is. (He cites "the Internet" as a comparative example.) It's why the artist named his series "Kudzilla." While beginning the project, he remembers thinking about a fictional monster arising from the natural system, a mythological and conceptual entity evoking irrational fear—a lot like Godzilla. "Rather than trying to talk about the plant, we're talking about all of our fears about overgrowth, lack of control, maybe something foreign entering in your lives, that you don't know what it is," said Brown.

That very framing of the vine can be found if we travel backwards in time. In 1968, *The Kudzu: Subterranean News from the Heart of Ole' Dixie* launched in Jackson, Mississippi, where it ran for four years, producing 6,000 copies per issue, with an estimated circulation of 1,200 per issue. Founded by former and current students at Millsaps College through the Mississippi Student News Project, as well as members of the Southern Students Organizing Committee, and predominantly distributed to locals and college students, the publication was funded by founding and early editors—David Doggett, Cassell Carpenter, and Everett Long—and operated out of an apartment in the city. A 1968 issue of the paper billed itself as "providing the only unbiased local reporting of the civil rights movement, peace movement, and the counter-culture." Covering cases of discrimination against Black people and women in Mississippi, as well as criticism of the American involvement in the Vietnam War, *The Kudzu* was said to be one of the very first underground newspapers providing evidence of a counterculture movement in the Deep South. Its staff endured harassment and reported abuse at the hands of local police; Doggett wrote in an issue of the paper published in October of 1970 that "two members of staff have been beaten by cops in the past five months." A month later, another story was printed alleging that "police planted dope, destroyed private property, stole files and records, and arrested eight in one fell swoop." In 1972, just four years after launching, *The Kudzu* printed its last copy before closing shop.

The era was rife with mentions of kudzu in Western pop culture. Artists, writers, and musicians scrambled to reference the tenacious vine in their work. In May of 1963, Georgia poet James Dickey published a sensational poem about it, while at *Harper's Magazine* from 1963 to 1971, Mississippi-born editor Willie Morris labeled it as "sinister" in his musings about the South. Arkansas native Johnny Cash sang about it in his 1979 cover of the ballad, "The L&N Don't Stop Here Anymore," and Jimmy Buffett, who came from Mississippi and was raised in Alabama, snuck in a lyric about a crown made of kudzu in his 1994 release "Love in the Library." Other notable musical references include Texas singer Rodney Crowell's 1994 song "The Rose of Memphis," indie folk band The Mountain Goats' 1995 tune "Stars Fell on Alabama," and country rock group Alabama's 2001 single "When It All Goes South"; all included a nod to the prolific vine. Marjie Short's *Kudzu,* an Oscar-nominated 1977 short film, featured interviews with then-president Jimmy Carter, novelist James Dickey, and members of the southern rock group The Kudzu Band. Sheets of the vine blanketed trees in the background of Tom McLoughlin's 1986 horror flick, *Friday the 13th Part VI: Jason Lives.* And in 1996, filmmaker Max Shores released *The Amazing Story of Kudzu,* a documentary about the vine that is still referenced in modern podcasts and media unpacking its abundant presence. That same year, Ron Harrist published "Kudzu: Dixie's Creeping Death," an Associated Press news article that describes the plant as "a nightmare that at best can only be controlled."

Fast forward to today, and the enduring influence of the plant can still be abundantly found in unexpected pockets across the American South. Hanging in the Mississippi Museum of Art is Carroll Cloar's *Kudzu*, a 1976 painting that depicts a young girl walking on a road surrounded by tsunami-sized walls of the plant. Over in Decatur, Georgia, a 25,000-square-foot antique store goes by the name of Kudzu Antiques. In Huntsville, Alabama, a video production company calls itself Kudzu Productions. The vine has also made recent cinematic appearances. *Under the Kudzu,* Claudia Stack's 2015 documentary, chronicled the history of two Rosenwald schools in Pender, North Carolina, built during the segregation era, and *Kudzu Zombies* is a low-budget thriller released in 2017 where a kudzu herbicide turns meat-eaters into killer zombies. "The Carrot in the Kudzu" is the name of a 2014 episode of the Fox Network crime drama television series *Bones.* In it, a man's corpse is discovered in a canopy of kudzu, which is depicted as a flesh-eating plant that a team of forensic anthropologists have to contend with in their investigation into the human remains. You can even catch a reference to the weed on "Save My Place," a 2023 episode of the Netflix hit show *Sweet Magnolias,* where the series protagonist complains about a woman who is "just like kudzu, creeping all over everyone, choking the life out of them."

It still shows up sporadically on the country music scene, too. Darius Rucker sings about it in his 2013 song "Take Me Home," "Gemini" is a 2015 track by Alabama Shakes that references the plant, and Florida Georgia Line croons about being metaphorically tangled up in the vine in their 2020 song, "Countryside." Grammy award–winning singer-songwriter Amanda Shires also has a 2011 track named after the weed. "Kudzu" is an ode to love and what blooms in the countryside.

As for iconic citations in literature, internationally renowned author of *The Handmaid's Tale* Margaret Atwood has contributed to cultural depictions of the ever-growing southern insignia; her *MaddAddam* trilogy features genetically engineered adults who feed on it, as well as enormous "kudzu moths" who flutter through a dystopian world. Atwood's not alone—the vine has been the focus of scores of books. In 2000, Hal Crowther's *Cathedral of Kudzu* hit bookshelves, 2007 saw the release of Cathy Holton's *Revenge of the Kudzu Debutantes,* followed by *The Secret Lives of the Kudzu Debutantes,* Jennifer Lynn McCarthy published *Kudzu Salad: Short Stories from Florida's Forgotten Coast* in 2016, and J. D. Wilkes released *The Vine that Ate the South* in 2017. More recently, Stephanie G. Sewell published the thriller *Beneath the Kudzu* in 2021, Julius Thompson's pre-apocalyptic horror novel *Killer Kudzu* was published early in 2022, Tristan Tuttle's poetry collection *A Kudzu Vine of Blood and Bone* followed soon afterward, and Mimi Herman's historical fiction novel *The Kudzu Queen* hit bookshelves in 2023.

Sports teams and competitions have been named after it: Carolina Kudzu is a Raleigh, North Carolina, sandlot baseball team that has been playing since 2018, the Kudzu Classic Championships is a Georgia volleyball tournament launched in 2023, while the Kudzu Classic Youth Soccer Tournament is an annual affair in Mississippi. Ever since 1998, ultimate frisbee teams across the South have congregated in North Carolina for a three-day competition known as the Carolina Kudzu Coed Classic. Events and organizations representing every interest and affiliation also can be found bearing the unmistakable brand. In Atlanta, Georgia, Camp Kudzu is a camp for diabetic youth that has been open since 1999. In North Carolina, the Kudzu Warriors are a volunteer group working to eradicate populations of the weed that have appeared both inside and along the edges of the Norman Wilder Forest, a 185-acre nature preserve in Polk County. A couple dozen miles north, Chimney Rock State Park hosted "Get Krazy with Kudzu," in 2018, where visitors got to taste deep-fried or jellied kudzu and experimented with making baskets out of its vines. And every St. Patrick's Day since 2004, a group of women known as the Kudzu Queens don curly red wigs, tiaras, and bountiful green feather boas for Columbia, South Carolina's annual St. Pats in Five Points parade, raising funds for women's charities, including shelters and support for domestic violence survivors.

The name has also manifested in other buzzworthy ways. In 2011, multiple news organizations covered the viral story of how people claimed a patch of kudzu clinging to a utility pole outside of Kinston, North Carolina, resembled Jesus on the cross. Race car driver Jim Downing designed and drove the Kudzu DG-3, a sports prototype race car, which competed from 1993 to 2000. Chris Totten, a video game designer who also teaches at Ohio State University, is working on *Kudzu,* a game in development touted as a "a *Zelda: Link's Awakening*–inspired nonlinear adventure where a gardener is trying to save his mentor from a world-eating invasive plant." First premiering in the summer of 2020, you can regularly tune in to new episodes of *Kudzu Killers: Homicide and Sweet Tea,* a true crime podcast that bills itself as "a couple of honest-to-goodness southern gals discussing compelling tales of southern-style homicide." And Chris Lindland, a self-dubbed "monster botanist" and former student at the University of North Carolina at Chapel Hill, made the news in 2013 when he launched Kudzilla—a Kickstarter looking to raise $50,000 to build a 40-foot Godzilla structure made entirely out of the plant. Lindland was ultimately unsuccessful, raising just under $12,000, per his fundraising site.

"There's this hate and fear and vitriol that's sprayed on this plant. It makes me laugh," said Brown, who thinks the pervasive stereotype is unjust and misses the point entirely. The artist challenges that through his visualizations that sensationalize societal perceptions of the plant, hoping people can recognize that it has become a scapegoat for a deeper subtext. "We're extremely bad at conceptualizing, analyzing and clarifying what it is, what it does, whether it's negative or positive in our lives," he observed. There are a number of parallels to Godzilla, according to Brown. In his eyes, Godzilla is a depiction of extreme weather events, representative of the earthquakes and tsunamis that threaten Japan. A fictional primordial monster awakened by nuclear radiation, it is also most commonly recognized as a metaphor for the desolation and fear wrought by the 1945 atomic bombings of Hiroshima and Nagasaki by the US during World War II. It is a monstrous entity that arises from the ocean and descends from the skies, destroying cities and livelihoods without warning. It is a fanciful personification of the natural world, a way of thinking that has become entangled in how Americans attach narratives to a weed. It is, in many ways, a message—one that gestures to humankind's capacity for horror.

"We heap on the same kind of ideological baggage of Godzilla on kudzu."

13

MAN OF THE HOUR

Like us, plants are protagonists in their own stories. Like us, they rely on oxygen to survive. Like us, they are understood through the intricacies and influences of language. Perceptions left at the doorstep of mercy, a smattering of words deciding who we are.

It's those very complex and powerful nuances embedded within linguistics that compel the work of researchers like Derek Alderman. A professor at the University of Tennessee at Knoxville, Alderman is a cultural and historical geographer investigating how humans talk and make sense of our profoundly confusing world, documenting the relationships between people and their environments. At first glance, the fifty-four-year-old strikes an imposing figure, outfitted with the unspoken airs of someone who has been an expert in their field for some time. A shock of red hair adorns his head and his face, glasses perched on a studious nose. Any degree of intimidation is dissipated, however, once he starts to speak. In a smooth southern drawl, Alderman immediately cracks a joke, putting his conversation counterpart at ease. "If you hear music in the background, it's not a disco," he says, by way of introduction. We're on a Zoom, with Alderman speaking to me from a coffee shop in Knoxville, where he's stopped by to chat in the middle of a busy day of bustling around the city for work. The bulk of his work focuses on civil and social justice—although what he's most known for is kudzu.

His journey with the vine dates back to the mid-1990s, when Alderman was teaching undergraduate students at the University of Georgia. Wanting to try something new that would engage his classes with a topic sure to interest them, the geographer landed on lessons around kudzu. "I'm teaching to large lecture halls of southern students," he recalled. "How can I teach about the environment, resources, society space, in ways that will resonate with them?" By then, he had already read the work of John Winberry and David M. Jones, who were among the first academics to publish a journal article on the cultural geography of kudzu, which came in the form of a 1973 paper in *Southeastern Geographer*. After incorporating the work into his curriculum, the rest, as they say, was history. "I started looking at kudzu and noticing how many people were using the word kudzu, the image, the idea of kudzu in their everyday language," he said.

It's become a distinctive part of southern vernacular—often applied as a label for something that's a nuisance, overwhelming, or unwanted. Alderman has published numerous works on the vine and its role in southern culture, including journal articles investigating the discourse surrounding an anti-kudzu law passed in Missouri in 2001 and a book chapter in 2018 detailing the entangled relationships formed by people and places because of the plant. But what he's most known for within the body of literature on kudzu is his work spotlighting the untold account of one media figurehead from the twentieth century behind its climb to infamy. Alderman is one of a select few modern researchers to ever publish an article chronicling the biographical background of the man who helped make kudzu a southern household name in the 1900s—Channing Cope himself. Scouring archives and old newspaper clippings, it was the early 2000s when he set out to unearth everything he could about a character no more than a mere footnote in most literature on kudzu. "My research is often about, and I don't mean this offensively, but turning chicken shit into chicken salad.[1] I take the traces and remains of things that don't seem to belong together and find out how they fit together," he explained.

The idea to investigate Cope emerged after reading what the existing body of English literature had to say about kudzu's spread. "If you look at some of the early writers, they reference Cope very quickly, like some sort of joke," he

[1]. The expression "turning chicken shit into a chicken salad" means to turn something terrible into something of value. It was once famously said by US President Lyndon B. Johnson. See Cleve R. Wootson Jr., "A History of White House Profanity—and One Cursing Presidential Parrot," *Washington Post,* January 12, 2018.

said. Interviewing some of those authorities led Alderman to understand that most people chalked Cope up as "some sort of weirdo." He thought to himself: "There's got to be something more." At the time he was living in North Carolina, and this dive into Cope's past would send him across the South, scouring the Special Collections at the University of Georgia and the Atlanta History Center. At one point, he "dragged" his family out to Covington, Georgia, to try to visit what remained of Cope's farm. While in Covington, Alderman connected with Phil Cohen, Cope's longtime friend who was then still alive, who provided him with new insight into one prolific man's backstory.

It was 2004 when the cultural geographer published a paper in *Geographical Review* detailing Cope's legacy, compiling evidence on the role he played in romanticizing kudzu in the South. Twenty years later, that piece remains one of the most extensive resources on Cope—and the case for one media figurehead's lasting influence on American kudzu expansion. "I just thought, 'We've gotta get past judgments and take him for his word.' Not in the sense of believing it, because he said some outlandish things about kudzu, but taking him for his word in the sense that he had legitimacy. And that's what prompted me to go deeper into understanding who he was," noted Alderman.

Much as he did everything else, Cope emerged on the kudzu scene with panache, a farmer-turned-journalist who helped amplify the Soil Conservation Service's stymied introduction of kudzu to farmers across the country. Before Cope got involved, the wave of intrigue attached to the vine as an agricultural staple came and went, making no more of a splash than the next topical plant of the time. That all changed in the 1920s and the 1930s, which was right around when the federal government had decidedly thrown its weight behind marketing and incentivizing planting the weed. In it, they saw a glittering promise: an end to soil erosion, a major economic setback to the States, as they knew it. By then, kudzu translated to quick money. *Business Week* even designated it "cash on the vine." Farmers were offered up to eight dollars an acre to plant kudzu, as federal agencies intent on peddling it ramped up incentivization in a hopeful bid for the vine to be planted across the country. There was a special focus on the South, where a long-running period of intensive cotton production left farmland disintegrated and depleted by overuse.

One of the voices driving this widespread impetus to mobilize and assimilate kudzu was Cope. A 1949 *Time* article vividly brings to life a portrait of the man in question. In that piece, he is described as a "(205 lb.) genial South-erner" who "rolled reluctantly out of bed, downed a cup of coffee laced with bourbon, pulled on a shapeless seersucker suit, and started reading aloud to warm up his vocal chords." That same article mentions the then-recent success of Cope's debut book, *Front Porch Farmer,* which sold more than 80,000 copies, was Atlanta's No. 1 nonfiction bestseller for a spell, and was said to have been placed on the agricultural curriculum of at least four universities. Alderman himself once described Cope as someone who "played a significant role in making kudzu part of the working vocabulary of southerners in the early twentieth century." In a time when computers were as much of an in-comprehensible dream as a weed that could smother an entire region, what we know about how Cope got to that point of power is predictably limited. He was born in 1894 in Louisville, Kentucky, to William Cassander Cope and Julia Evans Cope. William was a Baptist minister, editor, and college professor. Julia was a schoolteacher and founded the Parental Home System of Kentucky for orphaned children. Their sons were Channing and Willard.

Born in Atlanta, Georgia, the Cope matriarch was an active member of the Baptist church, the Order of the Eastern Star, the Kentucky Club, and the At-lanta Writers Club. She also greatly contributed toward creating the Authors' Grove—dubbed a "living memorial for writers"—in Atlanta's Piedmont Park. The Special Collections Library at the University of Georgia has a handful of mementos preserved for the Cope family, including a single letter from Julia to the Georgia General Assembly trying to abolish capital punishment in the state. Also included in their collection is a photocopy of her obituary from the *Atlanta Constitution,* noting her death on March 6, 1929. Historical records suggest that long before his mother's passing, Channing Cope moved to Georgia and landed a job working as an agricultural manager at a utilities company. In 1927, he bought his own farm—referred to rather fondly as "Yel-low River Farm"—where he began experimenting with agricultural innova-tions like kudzu, which, by then, was teetering at the cusp of its route to mass adoption throughout the Deep South. Throughout that period of discovery, he would marry three times. His first wife was Nellie Marquardt Cope, who died on October 21, 1929, just a few months after Cope's mother, according to genealogical records. His second wife was Helen Athleen Rohrer, whom he married the following year. His third and last wife was Ruth Gentry Cope, who would outlive her late husband by more than two decades. "A farmer is lost without an understanding wife beside him," penned Cope in *Front Porch Farmer*. Not much else is known about Nellie, Athleen or "Ruthie"; nor is it clear if they played a part in Cope's hero-worship of kudzu.

In his prime, the decidedly public-facing man was a regular columnist at the *Atlanta Constitution,* which had a circulation of nearly 200,000, and a radio personality with a daily program on Atlanta WCON, which worked very closely with the newspaper. Every day, Cope would broadcast directly out of Yellow River, his 700-acre farm in Covington, situated just forty miles outside of Atlanta. A 1948 *Broadcasting Magazine* issue praised Cope as a "pioneer in soil rehabilitation" whose "early morning Monday through Friday broadcasts are just what the farmer ordered." That same article editorialized why he was such a hit, with the author speculating, "Half the people in Atlanta have a rural background making Mr. Cope's rural remarks equally popular among the city slickers." And a 2017 piece on kudzu in the now-defunct news website *The Awl* described his "flamboyant" radio presence as a "Howard Stern of the Depression." Based on these accounts, it's safe to suggest that Cope had an "It" factor, a brand of charisma that resonated with the thousands of listeners and readers who tuned in to his every word. Whether it was verses decrying alcoholism in his column, "Channing Cope's Almanac," or fervent passages about land erosion forever entombed in *Front Porch Farmer,* his body of published work makes one thing clear—Channing Cope was a persuasive, passionate man. He was also a powerful one, flaunting a sizable media platform. "He was a prolific newspaper columnist. He had his radio show. He had the ear of major environmental public policy people. He was a booster and a bit of a maverick," Alderman observed.

That's why, in the early 1930s, when Cope started raving to listeners about a Japanese plant that could be the end to all land erosion, people paid attention. He was clever about it, too—heralding the perennial from East Asia as a solution to the considerable agricultural production problems afflicting farmers across the South. Cultural geographer Alderman makes the case that the language Cope used to describe kudzu contributed to the broader cultural perception of the plant, citing examples in his research of how Cope characterized the South as in "severe crisis," posing kudzu as the be-all and end-all remedy to something plaguing many southern farms and families. "Kudzu isn't a vine, merely. Kudzu is the Lord's indulgent gift to Georgians," Cope once noted, praising the way it could grow quickly and was said to rapidly replenish tired, dehydrated soils. It wasn't long before the media personality became known as the "Kudzu Kid" or "Kudzu King." Cope was one of the primary forces driving the southern agricultural love affair with a plant with origins from elsewhere.

"When we think about invasives, and we think about propagating invasives, there's the material we're getting out there, and the species gets introduced, they get this cultivating, it gets promoted in a material, landscape way. But there also has to always be the cultural work of framing how people see it, how people relate to it," explained Alderman. Cope was very good at doing just that—he used an affinity for storytelling to paint portraits of the salvation one invasive plant offered. "I don't think kudzu would have taken off as it has if not for the boosterism of folks like Channing Cope," Alderman affirmed. "They weren't just planting the vine, they were also planting it in our hearts, planting it into our heads, that *this* could be the escape."

Other excerpts from Alderman's work pull the curtain back on how Cope presented the vine as a balm to soothe suffering soil in an expanse of land—simultaneously poetic and poignant. "Some merciful soul, in a spirit of compassion for the naked land . . . made the effort to clothe the lifeless area in a garment . . . of green," Cope wrote. "This was six years ago. Today, the area lies covered under a blanket of kudzu duff: soft, warm and healing." In another passage, the farmer-turned-media mogul evangelized the promise of kudzu. "There are hundreds of thousands of such acres awaiting the healing touch of the Miracle vine," Cope wrote. The "Kudzu King" had a knack for lyrical expression in his writing, one he employed to his advantage. "He was a master of resonance," noted Alderman. "He was able to use language that resonated with Georgians, southerners and Americans at the time." In the early 1900s, Georgia was a highly religious place, where faith was entrenched in everyday life. It was also in the middle of sweeping public health campaigns; mosquito populations and malaria were radical threats and drivers of premature death for rural American populations. Cope capitalized on that concern in the ways he humanized kudzu, doing something that was "revolutionary."

"It's not that he was equating kudzu with malaria," observed Alderman, "but it's this idea that he's talking about issues that resonate with people."

In his prose, Cope managed to knit together religion with health, a nexus of crucial topics that spoke to his readers, evoking vivid, relatable imagery in their minds as they learned about the alleged miraculous healing qualities in a swath of celestial kudzu. By then an ardent kudzu devotee, he went on to found the spindly weed's national fan club: the Kudzu Club of America, which reported twenty thousand members in 1943. That same Georgia-based initiative had a public mission of planting eight million acres of kudzu throughout the South. It held annual festivals and elected "Kudzu Queens" to proverbial, plant-ridden courts. In those days, the Kudzu Club of America worshiped a symbolic plant as much as modern-day followers admire today's deified celebrities. Cope was then, and has since, been largely credited with the widespread rush to plant kudzu, a frenzy leading to kudzu seeds being planted on public lands across the Southeast. Kristen Hinman reported in a 2017 *History* article that *Time* had deemed Cope the plant's "chief cultist."

It was on the kudzu-kissed soil of his farm that Cope took his last breath. In 1962, the sixty-eight-year-old's heart suddenly stopped as he stood only a few feet away from his front porch. Reports of his death suggest that overexertion from chasing after hormone-frenzied teens, neighbors who reportedly used his kudzu-ridden property as a make-out spot, turned out to be too much for his arteries. Thanks to the next Yellow River farmland owner, the world would later learn that the weed had engulfed much of the farm's seven hundred or so acres, with some accounts suggesting the plant had swallowed entire portions of the property. There's something almost paradoxical in the way the "Kudzu King" at last relinquished his throne, surrounded by the very plant he worshiped, bands of the vine there with him in his last moments of life. Years later, the handful of accounts chronicling his passing tend to point to the warped irony.

But Cope's kudzu-riddled legacy was experiencing fallout even before he died, as farmers, the agricultural industry, and even the US government began grappling with the consequences of the overzealous "miracle vine" movement spurred by the media tycoon. Despite that, Cope managed to remain relatively unscathed; his daily broadcast and column served as outlets for a deep-rooted plant passion up until the day he passed. "If it hadn't been for kudzu the miracle vine, all my land would have washed away. They used to say cotton is king in the South, but I say cotton isn't king anymore—kudzu is king," he once said with dogged flair. Looking back, one can't help but wonder: Was the man a misinformed conservationist, or merely a media figurehead seeking nothing but glory? Alderman believes that he was a bit of both, part environmentalist and part showman. Cope was a person who craved attention, but he was also convinced kudzu was good for farming and ecology. "For someone to invest so much time and money and energy and their self, their identity into something, I think he honestly believed that it was environmentally the right thing," said Alderman, adding that Cope's ambitions for fame were inseparable from his views about the vine. "Channing Cope's life reflects how much kudzu had become his identity, and his identity had become wrapped up in kudzu." Whatever his driving force may have been, Cope's promotion of the benefits of propagating kudzu would have him go down in history as an immensely admired agricultural preacher in the South. Be it prolific raconteur or fervent environmentalist, whether he truly believed kudzu was an extraordinary ingredient or just jumped on a bandwagon for a shot at glory, one thing is obvious—Channing Cope left behind a legacy twisted in kudzu. "It entangles itself in people's lives and their identities, their language, in ways that make it impossible to simplify and make it impossible for us to reduce down to just going back," mused Alderman. "In many ways, my own identity and life have become entangled with the vine."

As Cope's unofficial modern biographer, Alderman has also been drawn into the American kudzu narrative. It's a fitting legacy for a lifelong resident of the South. His mom hails from the area surrounding Savannah, Georgia, and his father in the southwestern part of the state, near Moultrie. Growing up, Alderman's dad worked in government jobs, which led the family to relocate to Florida, South Carolina, North Carolina, Virginia, and Tennessee, constantly moving across the region. Having spent his childhood in different pockets of the South, Alderman knew kudzu without ever really knowing kudzu. "I always saw kudzu. I always grew up around it. I didn't know its real story. I just saw it as part of the background," he said. His parents were born in the 1940s, when the vine was a regular fixture in American popular culture—especially in places like Georgia, where by then it was widely established. "It's something that they always grew up around and they saw," he said. His investigations into kudzu have resonated with them. "They've appreciated the fact that I've always tried to explain or understand what makes the South the South," he added. "What is southern culture about?"

Beyond paying tribute to Cope, Alderman has experimented with communicating how kudzu transitioned from just being a plant in the southern landscape to a fixture in everyday life. "It's become the lexicon of people's language. It has become a metaphor that people use to describe their larger reality," he explained. "It's about how people tell stories about themselves, about their world. And kudzu's become part of that shorthand that they use to talk about the world." Another of his journal articles, published in *Southeastern Geographer* in 2015, compiled pages of examples of how kudzu has been framed in American discourse as a metaphor—representing both the idea of uncontrollable growth and invasive species as an entity. This includes characterizations of the vine to describe the rapid-fire spread of the Internet, urban growth, and even government regulation. The article also focuses on how kudzu has been deployed as a way to identify the South, through the naming of businesses and streets. Forty-five streets in the US have kudzu in their name; the majority are concentrated in North Carolina (15), followed by Georgia (9), South Carolina (6), and Alabama (6). While these homages to the vine are located in the Southeast, Alderman found that the influence of kudzu as a namesake surpasses these regional borders—as of 2015, 108 businesses incorporate kudzu in their name, spanning 21 states, from California to New York. Still, the vast majority are located in the South—with 42 in Georgia, 13 in North Carolina, and 10 in South Carolina.

Although kudzu has been scorned historically, the modern naming of commercial entities after it—which Alderman reports is done with the purpose of appealing to an intended consumer base—demonstrates how a vine with origins in Asia has been reconstructed to represent the South. After all, naming a place or business after the weed demonstrates an attempt to assimilate it into a regional identity and repackage it as a symbol of that area's idiosyncrasies. "What began as an Asian vine has become, for some people, a southern one," said Alderman.

"I really believe that kudzu's story entangles itself."

14

ALIENS
IN AMERICA

Julius Thompson was born in Statham, Georgia, a small town located right off of Highway 29, dropped smack dab between Athens and Atlanta. The brick buildings, asphalt roads, and power lines of this town of three thousand are surrounded by a dense thicket and a distant towering line of trees. Thompson still lives there, in a part of the state where seeing kudzu on a daily commute is a given. He's an author, behind titles like *The Ghost of Atlanta* and *Phantoms of Rockwood,* who has most recently turned to creating fictional worlds built around kudzu as a way to share his thoughts on the omnipresent vine. In January of 2022, he self-published *Killer Kudzu,* a part horror, part sci-fi book about a carnivorous mutant kudzu vine that gobbles up people across Georgia. If you ask Thompson, he'll tell you his fiction draws on specific modern influences. "I [have] thought of kudzu as another form of COVID that spread. It is like spreading racism, how it spreads. You can think of it as spreading evil, or it could be spreading good, it can spread change," he said. "It can cause people to realize that the little petty things we have that are different, are the same."

At face value, the novel probes the havoc wreaked by a mutated flesh-eating kudzu vine, something that truly could be dredged from a nightmare. But it's far more than that. *Killer Kudzu* is a book of metaphors, one that poignantly reflects on what it means to be a Black man living in the Deep South, a swath of the country where the majority of registered voters align with conservative ideologies and where neighbors can be spotted proudly waving Confederate flags, public remnants of a region's fundamentally racist roots. The seventy-four-year-old worked a lot of his own struggles into the narrative. Drawing on his experiences as a Black man coming of age in areas where Jim Crow laws were enforced, Thompson has experienced a lifetime on the receiving end of both subtle and overt racialized prejudice. "We were on segregated buses," he recalled. "There was the rope that says, 'Colored,' in the back and we had to stay behind the rope. And when we stopped off, you could only go to certain bathrooms. All that stuff is in there."

One of the most memorable characters in *Killer Kudzu* is a county commissioner, who is depicted in the text as a "racist, misogynistic white supremacist" who cheats his way into public office. Inspiration for the villainous character, Thompson noted, came from former US president Donald Trump. "I needed an idea, something to get started with," he said, chuckling, before detailing the ways the book's characters, like the Trump-esque antagonist, are worked into the plot as an intentional nod to incessant issues of racial inequality in the South and the social divides that have long persisted within parts of Georgia. The novel also features appearances of members of the Ku Klux Klan. (Spoiler alert: Many of these characters meet an untimely end at the hands of one personified, ravenous vine.) Although horror-themed on the surface, Thompson's use of kudzu in his work is representative of a larger, nuanced message. He sees it as a unifying force, a symbolic approach to how something that appears to be a curse can make strides toward a society centered on civil justice. In his story, the weed acts as a metaphor, a product of human creation that starts off as a foe but ends up bringing a racially divided community together. "That's why when I had the kudzu kill people, I had it kill everything. Black, white, animals, anything. It was a way of creating something bigger, that maybe can pull people together," he said. "People had to either work together or become kudzu food."

The mutated plant's senseless and random attacks purge an area of divisive thinking and hateful ideology while the community's unification against a common foe forges a bridge of peace for the people living there. "We have bigger problems to worry about. And the kudzu symbolizes bigger problems to worry about than just whether someone's Black, someone's white, or someone's gay. We have bigger issues to deal with," he noted. Racism and climate change and how the two compound in disproportionately inequitable ways are among these plights that deserve the bulk of our attention, according to Thompson. "And this is why, if we don't watch out, these bigger issues are gonna take us over."

Just like, all those years ago, when one vine took over the South. Not only is it symbolic in a negative light, but it moonlights as a force for good. "Kudzu is something that really symbolizes the South, because it shows the connectivity to who we are, because it connects us all," he observed. "I look out and I see it in the whole area. You can't tell whether that is a Black area or white area. All you can tell is it is a kudzu area."

Elsewhere, orange-brown snails glide and yellow-and-black-bellied spiders weave webs of silk across a canvas of dark, glistening soil in a gallery in Frankfurt, Germany. There, in the summer of 2020, artist and poet Precious Okoyomon debuted *Earthseed*, an exhibition that centers on a room covered in topsoil, which housed a vibrant corridor of kudzu. That spring, Okoyomon and the team at the Museum für Moderne Kunst's Zollamt gallery had set the scene—laying the topsoil and scattering kudzu roots that Coco Romack of the *New York Times Style Magazine* reported were largely left to their own devices to grow. Okoyomon was harvesting kudzu precisely for the installation, using the room that would months later be overflowing with sheets of the weed, and the medley of critters and creatures that had found a home in it, as a living artistic metaphor, a vessel of meaning commenting on the cultural response toward kudzu in the US. Motionless in the field of lush vines are human-like sculptures made of wool, figures that are both emerging from the mass of greenery and succumbing to it. A booklet that ran with the exhibition perfectly encapsulates Okoyomon's powerful meanings within *Earthseed,* which was inspired by Octavia E. Butler's novels *Parable of the Sower* and *Parable of the Talents*—where readers are taught that the seeds of the earth can be moved and no matter the manner or destination, they will find a way to thrive. "The story of kudzu is written in the same language that objectifies both human beings and nature and produces violence, oppression and individuation," wrote Okoyomon and coauthor Hannah Black. "Like *Blackness* itself, Kudzu is both indispensable to and irreconcilable with Western civilization."

This kind of framing of the "vine that engulfed the South" as a symbol of the relentless persecution and unstoppable power of marginalized communities nationwide can be followed back through time. In an article for the *New*

York Times published in 1973, famed American author Alice Walker wrote, "In Mississippi (as in the rest of America) racism is like that local creeping kudzu vine that swallows whole forests and abandoned houses; if you don't keep pulling up the roots it will grow back faster than you can destroy it." For a 1998 study published in *Sociological Spectrum,* sociologists Kathleen S. Lowney and Joel Best amassed examples of how the vine has been a vehicle for race relations, both in the context that Black farmers were once blamed for propagating the vine after it fell from grace in the eyes of the American public and that the plant's Japanese origin was being used to fuel anti-Asian sentiment. It has even been used symbolically to describe the destruction caused by Western European colonialism on Cherokee culture in a 2021 essay, "Cultural Kudzu," written by Heath R. Robertson and published in the *Journal of Thought.*

For more than a century, kudzu has been caught in a racially charged sea of rhetoric. As a plant native to East Asia, the weed has most commonly become a masked medium allowing Americans to harbor xenophobic sentiment toward their neighbors of Asian descent—and more broadly, immigrants as a whole. "In many ways, kudzu provided people a real convenient way of othering and 'exotic-sizing' Asia, particularly that part of the world," noted Derek Alderman. He remembers reading newspaper accounts about the vine from the 1980s that connected the growing corporate influence of Japan on America to the use of the vine, observing a correlation in how kudzu was another method to "other" Japanese culture during that period. "One thing that I discovered, I think later in my career than I should have, was that although most people talk about that very consequential moment [in which Japan brought kudzu] to the American shores, and that no doubt happened, there was a lot of Trans-Pacific circulation that went on between America, into Japan and back, and that is just as complicit [in the vine's introduction]," he observed.

It was in the 1940s and 1950s that a shift in American perception of kudzu began to unfold, as southern farmers started to speak up against the plant that they had once deified. This happened just as Japanese-American relations hit their lowest point. American rhetoric around kudzu, a plant with roots in East Asia, really began to morph during the period marred by World War II and the fateful attack on Pearl Harbor. On December 7, 1941, Japan, one of Germany's largest allies in the war, launched a surprise military strike on the US naval base at Pearl Harbor, near Honolulu, Hawai'i. The strike killed 2,403 Americans and wounded 1,178. Later ruled in the Tokyo Trials to be a war crime, it spurred a changing tide in the way Japanese citizens, Japanese immigrants, as well as those with East Asian ancestry, were viewed in the US. In response, a wave of xenophobia aimed at Asians gripped the country. The federal government launched Japanese internment camps, which President Roosevelt himself called "concentration camps," where at least 110,000 people of Asian descent were incarcerated. Over half of those forced to live in these prison-like enclosures were US citizens, who had either immigrated from Asia or had Asian ancestry. It is indisputable that this period inflamed anti-immigration rhetoric, something that had already begun taking hold in the former Confederate States of America and would endure into the future as intrinsic to a politically conservative regional subculture.

More than eighty years after the fallout of Pearl Harbor saw Asian Americans thrown in jail because of their perceived race, xenophobic sentiment persists. A swelling tidal wave of anti-Asian rhetoric has been fueled by systemic racism and a rise in white supremacists taking public office, divisive figureheads supported by a growing portion of the public that share unforgiving views on immigration and deploy baseless fear tactics to depict Asians as "others," unwelcome beings from nations crawling with Western enemies. All of this bigotry is also buoyed by America's conflicting power struggle with China. Tension has heightened as a result of theories about COVID-19 originating in a wet market in Wuhan, China, which has correlated with a rise in targeted anti-Asian hate crimes nationwide, perpetrated by a majority of white attackers. Between March 2020 and March 2022, more than 11,400 hate incidents against Asian Americans in the United States have been reported, which "signaled a persistent rise in harassment, verbal abuse and hate speech that have plagued Asian communities since the beginning of the Covid-19 pandemic," according to Edwin Rios at the *Guardian*. In early 2023, news of a Chinese "spy balloon" even dominated major US news headlines.

This vendetta is not leveled solely against those Americans bearing Asian ancestry. It is a piece of a bigger picture, one that reveals an overarching condemnation of anyone in the US who originally hails from other places. It is a damning piece of evidence of something that has long already unfurled its roots across the West. And we see it when we look at the way unwanted plants, and unwanted people, are framed in everyday dialogue. A 2010 paper in *South-*

eastern Geographer analyzed texts published between 1991 and 2006 related to an anti-kudzu law in Missouri. Authors Anna Eskridge and Derek Alderman found "alien invader" was one of the three most prominent discourses that framed the vine. "If you think about it, this discourse resonates with a lot of Americans," Alderman told me in an interview. "We're defensive about other countries, and their influence on us." For a little over a century, "alien invader" has been used as a common phrase in entertainment, media coverage, legislation, and literature produced by state and federal agencies about kudzu. But this wording is not just used to describe a plant that most people want gone. For centuries, it has been exercised to signify people.

Alien invader. Submerged in those twelve letters lies an entrenched form of prejudice, an atmosphere of otherness that segregates the vine and other invasives from the thousands of other plants thriving in the American South. Like the militarized language often ascribed to living things with origins in other countries, in certain contexts, phrases like these can be brandished as a weapon—effectively intended to minimize marginalized cultures and communities in the United States. A place where "alien" has been used to describe outsiders since 1790.[1] And what a word to use—one that carries with it a long, haunted legacy, with roots that trace back to the fourteenth century. It is not just wielded as a term to describe a person who is not of a particular group or place, or an extraterrestrial, a creature that is decidedly not human, but also used to assign a quality to something that is "differing in nature or character typically to the point of incompatibility," as defined by *Merriam-Webster.*

According to attorney and policy advisor Sarah Rich, "you see fights over language all the time" in the immigration space. "There's different schools of thought on the extent to which language shapes our thinking, that locks us into certain ways of thinking," said Rich. Based in Atlanta, Georgia, she is a senior supervising attorney at the Southern Poverty Law Center's Immigrant Justice Project. Throughout her career, she has worked at the intersection of immigration, employment, and civil rights. "Obviously, the use of language

1. In 1790, during the Washington administration, Congress passed the country's first Naturalization Act, which limited naturalization only to noncitizens who were "free white persons." They were classified as "alien." Back then, non-white immigrants were not welcome to pursue US citizenship. See Andrew Glass, "U.S. Enacts First Immigration Law, March 26, 1790," *Politico,* March 26, 2012.

is a key political tool to get people to come up, to have quick associations between the term you choose to use and a whole set of values or a whole set of policy outcomes that you want to get them thinking about," she noted. A trend in modern American immigration rhetoric today is the language around the "Great Replacement Theory"—an abhorrent concept that a shadowy cabal of elites, often coded as "Jews and Democrats," are bringing in or encouraging migration of communities of color in the United States in order to end white dominance and the white demographic majority, cementing the left's control over politics. "In this theory, all Black and brown people will inevitably vote Democrat, even though no one has voting rights when they initially come to the country," said Rich.

This alarmingly racist, antisemitic ideology was invoked by the young white supremacist who murdered ten Black people and injured three others in a Buffalo, New York, grocery store on May 14, 2022, live-streaming his brutal killing spree on Twitch. The theory was linked to the far-right white gunman who, on August 3, 2019, walked into a Walmart in El Paso, Texas, where he shot and killed twenty-three people, wounding dozens, which the Federal Bureau of Investigation has since classified as domestic terrorism. The shooter later admitted to authorities he was targeting Latinos in a horrific hate crime. On October 27, 2018, eleven people were shot to death during worship services at a Pittsburgh synagogue by an antisemitic white gunman who targeted Jewish people, who he said were "committing genocide to his people." It isn't just isolated to the US, but being weaponized across the world. On March 15, 2019, two terrorist attacks were carried out by a right-wing white fascist who targeted Muslims praying at mosques in Christchurch, New Zealand, robbing fifty-one people of their lives and injuring many, all while live-streaming it on Facebook. A manifesto produced by the killer later revealed how his obsession with the "Great Replacement" fueled the massacre.

"It's clearly encouraging stochastic violence among a group of radicalized white young men. So that is a very obvious example of using language to create a threat narrative," said Rich. "Literally an invasion narrative." "Anti-alien" concepts like these are propagated every day in discreet, radicalized corners of the Internet, on sites like 4Chan, 8Chan, and Gab, where white supremacists congregate, as well as widely publicized spaces like Fox News, where a 2022 investigation published in the *New York Times* found that former Fox prime-time host Tucker Carlson mentioned the idea behind the "Great Replacement Theory"—that immigration was being purposefully exploited to change the demographics and electorate of the US—in more than four hundred episodes. Cable networks like Fox that broadcast misinformation to millions of viewers are among the first public places Rich points to as an example of conduits of discriminatory language about immigration. "We hear a lot of invasion rhetoric, as well, from the right," she noted. "It's always easy to paint an out-group as a threat to the in-group, but the ways that they defined the in-group and out-group are explicitly racialized." Rich is now seeing members of the alt-right attempting to boost anti-immigration rhetoric through the war in Ukraine—a tenuous connection riddled with inaccuracies. This weaponized framing includes messaging like, "Russia invaded Ukraine, but we're being invaded every day because of the open borders policy." The attorney doesn't hold back from deeming that "complete bullshit." "There are no open borders," Rich declared. "There were open borders for white people [until] 1921." That, she said, was the year Congress passed the Emergency Quota Act, which limited immigration by national origin and imposed quotas on European immigrants for the first time ever. Those same borders had long been shut for everyone else.

Exclusionary restrictions on immigration first emerged in the US more than a century ago. In her memoir, *Once I Was You,* journalist Maria Hinojosa detailed how a series of racist immigration laws targeting Asian immigrants arose after California's state government prohibited Chinese immigration in the early 1860s. Thousands of Chinese laborers—who tended to be cheaper than other laborers—had immigrated to the US after being recruited by US railroad tycoons trying to build a transcontinental railroad. Once they were built, those who had moved across continents for the work were no longer wanted. According to Hinojosa, former California governor Leland Stanford once addressed the state legislature in 1862 and said, "The settlement among us of an inferior race is to be discouraged, by every legitimate means. Asia, with her numberless millions, sends to our shores the dregs of her population."

Twenty years later would mark the first significant law ever to restrict US immigration—the Chinese Exclusion Act of 1882, banning Chinese laborers (loosely defined as "skilled" versus "unskilled") from moving to America for ten years and making it nearly impossible for non-laborers to enter the country. The legislation also added stipulations to Chinese residents already living in the US; they needed certification to reenter if they left the country, and they could no longer be granted citizenship by state and federal courts. This legislation became the base of the Geary Act, which was made permanent in 1902, and added further restrictions. It would remain the baseline until the 1920s, when the Immigration Act of 1924, or the Johnson-Reed Act, banned all migration from Asian nations. In 1943, when the US needed China's aid in World War II, the federal government repealed the ban but instituted a quota, only admitting a limited number of Chinese foreign nationals (an annual cap of 105 new immigrants). Restrictions on all other Asian immigrants remained in place until 1952, when legislation allowed a minimal number of visas for citizens from other Asian countries, including Filipinos and Indian immigrants. But it would be another thirteen years before the US Immigration and Nationality Act was signed, largely putting a halt to immigration restrictions connected to the country of origin—finally putting an end to Asian exclusion.

It was 1953 when kudzu, another "alien invader," was publicly removed from the USDA's list of permissible weeds, but widespread perception regarding the vine had already begun to sour the decade before. With the American public caught in the throes of a wave of anti-Japanese sentiment, could this timing have been a mere coincidence—or driven by something more sinister? Lest we forget, the word *alien* has also been used to classify non-native species since 1835, when English botanist John Henslow began applying this dichotomy to plants, using "native" to differentiate types of plants that did not belong to a "true" British flora. (Scientists first proposed using "alien" to refer to some plants in 1847.) Nearly two centuries later, the International Union for Conservation of Nature's Global Invasive Species Database relies on a system called the Environmental Impact Classification for Alien Taxa (EICAT)—which, quite literally, is self-described as "the classification of impacts of alien taxa on the environment."

If you ask Katrina Maggiulli, xenophobic messaging is still blatantly rife in the field of invasive species—and it is actively being imprinted on the malleable minds of children. A doctoral candidate pursuing environmental studies and policy at the University of Oregon, Maggiulli researches the ethical dimensions of conservation work. Her 2022 paper in *Environmental Education Research* analyzes the problematic ways invasives—like kudzu—are framed in American educational materials and literature. Maggiulli hasn't come across many others doing work on the ethical dimensions of education, especially in the context of invasive species, which was why she ended up working on the study. "It's a huge gap," she said. "People are talking about it, right? People are teaching invasive species everywhere. It's definitely one of the primary things people teach in environmental education. But I wouldn't say that they're teaching it the right way."

Ethical issues in education surrounding the topic became clear to Maggiulli while she was interning for the US Fish and Wildlife Service in 2013. There, she noticed the charged ways in which the people around her were engaging with invasive species. "I was noticing the ways in which the biologists would work with, on the one hand, the extreme tenderness, that they would treat all the native species they would work with," she said. "And then, I would say, an extreme, almost Jekyll-and-Hyde approach that they would take toward the invasive species that they would then engage with." She didn't like the way things were framed. "I felt like that was something ethically alarming to watch," she added. It wasn't long before Maggiulli became an environmental educator for kids. As she got more acquainted with the content and curriculum she was provided, she spotted several red flags. "It was very much structured, as like, 'These are the most evil creatures,' and it was really one-sided," she observed. "Wanted" posters criminalizing invasive species have been one of the most common things she's come across in her research—as well as an overemphasis on the "invader" and framing of these species as wrong and undesirable. "This was just very striking to me. That this was a species being morally framed, literally framed in a picture frame, as a criminal. And that was really, really interesting to me. Visually, morally," she said. As a teacher, she almost couldn't believe this was the messaging around these plants and animals that was being drilled into adolescent minds. "I was astounded that this is what they're being given," she said, "and I felt like I needed to be questioning this further."

One weighty example is the northern spotted owl, which is classified as a native, endangered owl in the Pacific Northwest. The emblematic species is specialized, in the sense that it can only live in old-growth forests and looks very similar to the barred owl—a species that is native to the eastern United States but has recently expanded its range across the country and is now in the western US, where the US Department of the Interior has determined that it's an invasive species. Scientists at the Forest and Rangeland Ecosystem Science

Center have also classified the barred owl as an active threat to the dwindling northern spotted owl population. "These extremely closely related birds are from the same genus, and they can actually interbreed," explained Maggiulli. Despite the barred owls' ability to survive in several ecosystems, the relatives have started competing for habitat in the Pacific Northwest, likely due to climate change and diminishing habitats shifting species ranges across the world.

In 2013, the US Fish and Wildlife Service (USFWS) decided to begin lethal removal of the barred owl. The agency released an environmental impact statement as part of an organized attempt to create population declines in the species in an official effort to eradicate the birds from the region. "Which was huge to do, right? Because these birds are very similar looking," she noted. "It's this one bird, on the one hand, that everybody absolutely loves because it's our native owl, and then a bird that looks very similar to it, but we're now killing it," she said. "One of them we've now deemed to be evil. And we're killing it, but it can interbreed with the other one. And when they identify one another, it's closely related enough to actually interbreed." A US federal appeals court ruled in favor of the USFWS in the court case *Friends of Animals v. U.S. Fish and Wildlife Service*, allowing the agency to move forward with its experimental plan to kill barred owls as a management strategy in 2022. "We should be teaching some of those ethical dynamics to kids," asserted Maggiulli, who believes invasive species management and conservation work are fundamentally ethical issues. "I feel like people forget about the fact that the most basic tenet of conservation is that we are identifying something of value in the world, that it's worth conserving, and therefore, it's an ethical matter."

It's impossible not to consider the cultural ramifications that come out of the ways invasive species and plants are framed, in everything from educational context to broader research approaches; oftentimes fraught, charged messaging, which is taught to the masses at young, impressionable ages. In one haunting case mentioned in her paper, the environmental educator came across a student project that had been widely shared on a teaching material site—and many times since reused—depicting a purple loosestrife flowering plant, as well as a lionfish, nutria,[2] and feral hogs, in a "Wanted" poster format with an American flag in the background and each species labeled as illegal

2. According to the USDA's National Invasive Species Information Center, the nutria is an invasive rodent in the US ("Nutria, n.d.," https://www.invasivespeciesinfo.gov/aquatic/fish-and-other-vertebrates/nutria).

immigrants. "It's really illustrative of this overlap and where the stakes are with this connection," she observed. "It just shows how easy it is to make that leap." The narrative in that exercise largely reinforces several damaging and deeply politicized ideologies. The first is the imagery of the undocumented immigrant depicted coming over the border, or being an "invader," to America, associated with an invasive species entering the country through a pile of firewood, and the added subtext linking immigrants with being wanted by the police and "unwelcome," to the way ecologists seek to eradicate invasive species. It tells a story of xenophobic anti-immigrant messaging being embedded into lessons about ecology and conservation. It also demonstrates that indoctrinated thinking can spring from unexpected places.

According to Maggiulli, the Wild West "Wanted" framing of invasives at the heart of this activity, a fervent criminalization of plants with origins from elsewhere, is "extremely common" and a regular practice for teachers educating youth on what invasive species and plants are. "Everyone's making those across those different groups. It makes any vilification of these different species . . . glommed onto these human communities," said Maggiulli. "Therefore, the stakes become really clear, that it's going to have impacts on human communities. It just can't not." She noted that such racially charged framing fits "so many" different invasive species. "There's a number of different species that were brought to be used for very specific purposes, which has really unfortunate parallels with other human immigrant communities, too. So it has unfortunate parallels with bringing communities in and then blaming human immigrant communities for other problems. There are so many really bad parallels with those kinds of narratives," she noted. "That's why those kinds of xenophobic narratives make so much sense. Because these histories repeat themselves, in many different ways. And it mirrors those narratives."

Discourse like this breeds other subtle messages, language associations that render invasive plants and animals as all-powerful, while native species are relegated to passive and weak parts of an ecosystem. Humans, in this framing, are glorified as heroes, godlike deities exerting the unique influence to "fight" the invaders and "protect" the plants and animals that "belong." What this type of imagery naturally dredges up is humans and natives cast as "good," while invasives are labeled as "bad," and intruding in new areas by their own, evil designs. So arise questions of accountability—as more often than not, the reasons behind invasive species and plants being abundantly introduced to ecosystems, overtaking native plants and animals and rewriting landscapes, comes down to the whims of humans. This is certainly the case with the "vine that ate the South." The perennial only became a pervasive challenge because of the federal government's vested interest—fueled by a misguided belief that the species could solve a significant economic issue plaguing a vital industry, without consideration of the ways it might react to nouveau environments. "You can't put guilt on a species," argued Maggiulli. "It's not the human. They don't have a judicial system; they are not part of this moral system in the way that we are. They are not the guilty party here. And so it is problematic to think of them as invaders that way."

A 2020 article by the nonprofit media outlet *Ensia* dissected this in the context of climate change. Journalist Jenny Morber wrote about the Indigenous perspective on migrating species belonging to Anishnaabe peoples. She made the point that some Indigenous communities look at plants as equal to humans and do not ascribe negative classifications to the arrival of new species to their lands. Morber also covered the work of Mark Urban, University of Connecticut ecology and evolutionary biology associate professor, who authored a study about the risks of climate-tracking species due to warming temperatures. That research supports the idea of modifying the use of tools like the EICAT. Rather than a universal standard to be applied blindly, there's also an argument for these kinds of classifications to be made on a case-by-case basis, according to Piper Wallingford. The climate resilience scientist led a 2020 study analyzing the adaptation of existing invasion risk assessment frameworks, which was published in *Nature Climate Change*. And a 2018 *Smithsonian* article by Amy Crawford challenged the use of inflammatory rhetoric—both in and out of the invasive species climate context. In "Why We Should Rethink How We Talk About 'Alien' Species," Crawford wrote about the social implications of using terminology such as "alien" and "exotic" to describe species—especially in the US, where immigration rhetoric refers to illegal human immigrants in the same way. Crawford interviewed Mark Davis, an ecologist whom she described as "notorious" for a 2011 comment published in *Nature* where he, along with eighteen fellow ecologists, argue against using militarized language in invasion ecology. "Don't Judge Species on Their Origins" cites the impracticality of "[restoring] ecosystems to some 'rightful' historical state."

Using examples where costly eradication attempts were misinformed or ineffective, Davis and his coauthors argued that "classifying biota according to their adherence to cultural standards of belonging, citizenship, fair play and morality does not advance our understanding of ecology." Tamarisk shrubs, invasive plants that the US attempted to wipe out over a span of seventy years, are one of the cases they underscored. Following the failed crusade, ecologists discovered that the plants use water at the same rate as native counterparts and are now the preferred home for the endangered southwestern willow flycatcher. "It is time for conservationists to focus much more on the functions of species, and much less on where they originated," the authors concluded.

Meanwhile, a developing collective of US-based scientists have been busy working to reframe the broader scientific community's approach to naming. One such initiative is the Entomological Society of America's "Better Common Names Project." Advocating for official changes to common species names that include "derogative terms," as well as those with "inappropriate geographic references" and those that "disregard" names given to insects by Indigenous communities, the group reclassified the "Asian giant hornet" to the "northern giant hornet" and the "gypsy moth" to the "spongy moth" in 2022. A surge of attention to the issue led the US Fish and Wildlife Service to officially halt references to the "Asian carp"—instead calling it the "invasive carp"—which the Illinois Department of Natural Resources took one step further in their ongoing petition to the FDA to rename the species "Copi" for "copious." On smaller scales, scientists across the country are beginning to openly acknowledge the cultural insensitivity packaged into many historic common plant names. Some, like the New Jersey Agricultural Experiment Station at Rutgers University, are using alternative names or removing references to a country of origin in their research publications.

But "alien" can still be found everywhere in the world of invasive ecology. The word is even incorporated into the Global Compendium of Weeds (GCW) and their grouping of species—*Pueraria lobata*, or kudzu, has been listed with seven different labels: "agricultural weed, casual alien, cultivation escape, environmental weed, naturalised, noxious weed, weed." According to the GCW, "casual aliens" are "species [that] appear with no direct (apparent) human assistance, survive, possibly set seed, but do not persist." Fourteen percent of all weeds in the listing carry that distinction, per a 2012 article in *Polish Botanical*

Society. Labeling plants and people as "alien" is only the beginning, which begs the question, Is it wise to even label these flora and fauna as "invasive," "non-native" or "native," to begin with? "They become guilty by name because the word invasive is—they are invaders, which ultimately, frames them as being guilty. And particularly, in that sense, because who is the guilty party? Well, humans, right? Because we're the ones that introduced them," Maggiulli told me. "That's a part of the narrative that people tend to forget."

There's no question that plants like kudzu are disruptive and dangerous when introduced to new ecosystems unprepared for their entry. They not only alter biodiversity but wreak havoc where they invade—scientists are in agreement that invasive species in general are one of the leading drivers behind the extinction of native flora and fauna. They're also costly. A 2023 report by the Intergovernmental Science-Policy Platform on Biodiversity and Ecosystem Services (IPBES) found that the global cost of "invasive alien species" exceeded $423 billion annually in 2019, with costs at least quadrupling every decade since 1970. Efforts to remove them from places they haven't traditionally lived in have also led to mixed results. From rats to goats, invasive mammals have been hunted, trapped, and poisoned in concerted efforts to wipe them out on islands where biodiversity had been reduced because of their presence. The ecological impacts of the eradication of invasive vertebrates on Palmyra and the Galápagos, some of the world's most remote islands, were scrutinized for a 2022 study in *Scientific Reports.* Researchers assessed the restoration rate of native flora and fauna in those areas, finding boosts in native populations post-extermination.

Take the Burmese python. The sometimes 200-pound snakes can be ecologically devastating in wonderlands they make their domains, like the Florida Everglades, where they are considered invasive. Possessing an appetite for native critters, mice and raccoons are among their most common meals. As they slither through ethereal swamplands, feasting on native fauna, the carnivorous reptiles alter the biodiversity at the heart of a resource of environmental riches. First spotted in 1979, the Burmese python has been connected to severely diminishing populations of raccoons, opossums, bobcats, marsh rabbits, cottontail rabbits, and foxes in Everglades National Park. A 2012 study in the *Proceedings of the National Academy of Sciences* indicated that several of these populations have declined between 99.3% and 87.5% since 1997. The evidence of its impact is all there—the bodies of the mammals belonging to these dwindling groups have often been found in the stomachs of Burmese pythons that once dwelled in the Everglades, like a serial killer caught red-handed. Except these snakes are not channeling the diabolical likes of Jeffrey Dahmer and Albert Fish as they gobble up a furry meal. They were transplanted out of their homes by humans decades ago and forced to adapt, to survive. (One example: More than 300,000 Burmese pythons were imported to the US between 1979 and 2009.) This doesn't mean the population in the Everglades shouldn't be culled. But perhaps we can rethink the way we talk about these issues in our storytelling, in our education and entertainment narratives.

Our use of language is powerful, still somehow that influence is often overlooked. There's no question that the words we use to describe and depict plants and animals like the Burmese python and kudzu carry radicalized, entrenched, and systemically problematic lines of thought and association. And when children are learning about invasives traveling across international borders in ways that criminalize both plants and humans, the lasting impact of the words we use for these classifications speaks to the barriers that persist in a country so deeply and pervasively divided. It's between these two spaces that the world of invasion ecology hangs in the balance. At a crossroads of sorts, some are caught between wanting to keep things exactly as they are and embracing the solutions in the unknown. A 2020 essay penned by Jenny Liou—"Am I an Invasive Species?"—in *High Country News* talks about this context through the lens of anti-Asian sentiment. "When popular media reports discuss *V. mandarinia* and the possibility of containment, it's in the unspoken but inevitable context of an Asiatic contagion that we failed to contain," wrote Liou of persistent narratives surrounding the "Asian" giant hornet. "The wasps present an alternate reality, an opportunity to track, corral and exterminate an Asian threat."[3] The same can be applied to kudzu. When rhetoric around a vine is so antagonistic, when a plant is militarized and demonized and painted as the enemy, it's simply impossible *not* to consider how a Western subconscious anti-Asian crusade is quietly, and loudly, fueling the sentiment. One emboldened by a rise in anti-Asian discourse and Asian American–targeted

---

3. The mention of *V. mandarinia*, or the Latin name for the "Asian" giant hornet, is misspelled in the original essay—as published in *High Country News*—and referenced with the correct spelling here.

hate crimes in the US, a trend that has disturbingly intensified with the onset of the COVID-19 pandemic.

All of these contexts overlap, like hopelessly tangled wires or roots at the heart of a mass of vines. Despite the implication that language can lead to cause and effect, the actions of the many speak louder than the words of the few. Maybe the real problem lies with the strict parameters we lay down. The ones we let rigidly guide us, and those around us, as we move forward in our misguided collective quest to find meaning in the undefinable. Maybe we've been posing the wrong questions about a weed at the forefront of southern culture and the crux of American invasive ecology. Instead of debating whether kudzu is the "vine that ate the South," could we be asking ourselves something new? Maybe there's more to learn from how we're seeking to understand and classify our biosphere, the world around us, and ourselves by looking through the lens of a species like kudzu. Over a hundred years of mistakes, meaning, and memory can be found within its bristly leaves, its sturdy roots, and its blossoms that glow with purplish hues. In the end, looking back at our history with it has a way of revealing the truth. Could we take a note from the pages of our past and rewrite the ecological narrative around kudzu?

It's a big ask, but it's also one that environmental educator Maggiulli sees the conservation community as ready to take on. Controversial or not, she believes in championing ideas that stir up a sense of disquiet like these— movements that force us to challenge our own implicit bias and the ways we may contribute to a ripple-effect problem, especially in the face of our warming world. As our climate continues to shift rapidly and extremely, so will the biodiversity and makeup of ecosystems everywhere. Amid all these changes, an opportunity to reflect, reenvision, and reimagine these institutionalized systems is likely to arise. "The kinds of ranges that various species are going to have to live within are going to be alternating, and so what's going to be 'invasive'? What's going to be 'native' and 'non-native'? That's going to be a difference, in the future, in general. We're entering changing times," said Maggiulli. As we see endangered "native" species in one region of the country become "invasive" in others, her colleagues working in conservation are going to find they will need to rethink the vocabulary they use. Maybe that will lead to the kind of shift she hopes for. "We might be talking about these different species in new ways," she posed. But if we don't classify plants from elsewhere, with roots in otherness, as "invaders," what should they be classified as? It's a question Maggiulli doesn't have the answer to just yet. Still, it's one she would like to see emerging in the world of conservation to be discussed. The possibilities are something she's thinking about, especially in the dead of another relentless summer, as waves of scorching heat and humidity beckon delicate magenta blossoms to bloom on strips of swinging kudzu.

"I feel like we're entering the future here. I feel like we're on the cusp of something new."

15

AM I
INVASIVE?

In plants, we can find similar tendencies that, on the surface, mirror the very elements that drive people. An innate, instinctual desire for survival. A gravitation toward community. A life that orbits around light. An invasive vine will conquer a field of thriving plants simply because this is how it knows to be. In that way, an unassuming weed demonstrates how entangled we are. Plants and people. Connected through a human-defined understanding of the world, a language that personifies every element of the universe around us. It is through that lens of language that we learn to accept that everything living is complex. Everything has a story. Even the creatures of the dark, sensationalized horrors that terrorize children's dreams, are more than they appear. We live in a world where stories have become an intangible form of currency, used to explain away the things we otherwise can't. Carefully constructed narratives are used to make sense of the senseless. Where mythical demons of the night face off against brave warriors defending mankind. Fictional characters serve as vessels for decoding the dichotomy, as we seek to explain how good people can do bad things. How we wrestle with the consequences of hurting and being hurt.

Like a bandage applied to a surface wound, dramatized tales are deployed as if they were weapons of mass destruction. But just like any artillery, they carry the capacity to impair. Fairy tales and proverbs whispered to lull little ones to sleep cement implicit stereotypes into young, impressionable minds. Passed down through generations, these bedtime sagas are woven with threads depicting age-old myths, embellished accounts of clear villains and handsome heroes and an infinite battle between right and wrong. Without being aware of it ourselves, these passages give rise to an indoctrination of children—fiction teaching us to believe clichéd truths about the parameters of the rigid system we're to grow up in. What we choose to share doubles as powerful messaging. From these childhood fantasies of evil dragons and perfect protagonists, we learn dangerous falsehoods. We grow up thinking that the monsters among us wear obvious, ugly masks. We mistakenly believe that if something is beautiful on the outside it must be inherently good within. We are taught that strangers are the only ones who wish us harm and that bad things only ever take place in the dark. Most of all, we are conditioned to think that the line between good and evil will never be blurry, that the right choice is always as vivid as a splash of startling bright watercolors on an empty canvas, as obvious as a storm cloud gathering in swirling, shadowy skies.

But outside of schoolbooks and nursery rhymes, the reality of the world in which we live teaches us grim truths. Sometimes, the horrifying monsters that walk among us wear perfect, pleasing faces. Sometimes, the people we know best are also those capable of inflicting the most pain. Sometimes, the scariest things take place in the light of day; nightmares play out in visible places surrounded by people who decide to look away, where none are willing to stop explicit acts born from hate. Life teaches us these lessons, and some of us are forced to learn earlier than others. By nature of our circumstances, of the ancestries we're born into, of the skins we cannot shed. We learn that the cruelty of others rarely requires the cover of shadows to unfold. We learn that people are often motivated by greed and a sense of glorified self-preservation. We learn that the space between good and bad is almost always blurred. And we learn that the things we end up fearing most are nothing like the vanquished creatures we once dreamed about. For the root of all of our problems is the most fathomless part within ourselves. It is the human need to conquer and to control the narrative at any cost. These lessons can be applied to the origin stories of everything that we've shunned and cast aside. Like the seeds once planted that flourished into a seemingly invincible vine that would go on to inspire urban legends of the South.

I set out intending to write a book about one plant's shifting cultural and scientific relationship with the humans and ecosystems around it. But with time, it turned into something else entirely. I embarked on this monumental task aiming to answer a straightforward question: What is it about our layered history with kudzu that makes one singular plant so riveting? Something so easy to humanize, so impossible not to. It wasn't the surface level of kudzu's story with the South that hooked, enticed, and drew me in. It was what the deeper meaning behind that tumultuous relationship could offer me. A child of immigrants. Someone born on this southern soil, but simultaneously so disconnected from what it represents. Perhaps this plant could offer me the peace, reconciliation, and acceptance I had spent my whole life, as a second-generation American, both running from and fighting for. It led me to question: How does my identity as an "other" define me?

Across the United States, the rhetoric around immigration has long been divisive, built on the back of systemic prejudice. Once, America represented a gateway to freedom. To the rest of the religiously oppressed world, it symbolized inclusivity, opportunity, and a chance at a better way of life. But when we read about the ways the nation ushered in people seeking salvation, fleeing from religious persecution, what is often glossed over is precisely who was ushered across the border. It wasn't people of color who were welcomed to the promised land. It was white Europeans who would go on to decry the rules and regulations of a new country, crafting amendments and a constitution that only favored the group they themselves belonged to. Still, the allure of the country was broadcast and its dream of freedom, a facade that was shrouding a nightmare for many, echoed like a siren call across the high seas.

There was a darker side to this land of possibility, a sordid truth simmering beneath the surface. Those who founded the nation brought with them promises of blood money—wealth acquired through a legacy of enslavement. For in the new order of America, only white men were allowed the liberty promised to all. Indigenous tribes were forced to relinquish their lands and communities, and were murdered in cold blood. Black men, women, and children were stolen from their homes and dehumanized as things to be purchased, sold, and owned. Women were considered empty vessels of motherhood and sexual gratification, unworthy of opinions and opportunities.

Centuries have since passed, and injustice still overwhelmingly persists today. We see it in the disproportionate slew of cases of police brutality targeting Black communities in the US, who are spurned by a largely ungoverned agency built on a foundation entrenched in racism and prejudice. We see it in the American incarceration system, an entity established to perpetuate the enslavement of nonwhite bodies. We see it amid the politicized war on abortion access, as men seek to limit and lock up a woman's right to choice, attacks that are rooted in white supremacy. We see it reflected in the makeup of our legislators, with white men holding 62% of all elected offices in the US, despite making up less than one-third of the population. We see it in the way that luck is praised as a virtue by most Americans, and a lack of it is condemned, all while the nation's long track record of discriminatory policies contributing to racially based economic barriers is rarely taken into account. And perhaps most tellingly, we see it in the way the South has struggled to stray far enough from its historic legacy of divisiveness; evidence that is found in the presence of Confederate flags waving proudly from porches strewn across the region, emblems of a heritage stitched together by threads of hatred. Banners standing tall in their prejudice, unmoving in the face of winds of justice, deaf to the cries for change. It's a terrifying tale, one nourished by seeds of mistrust purposefully scattered, planted, and nurtured by those who stood to gain. It's one I've felt the repercussions of firsthand.

I am a patchwork American, strung together by DNA strands from distant places. Meant to exist in the margins. These are my labels, my unshakable question marks. Belonging to no one, with nowhere to belong. I am still struggling to understand where I fall, where I fit. Am I invasive? Brought here by people who didn't know any better, who dreamed of more and ended up with much less. But this burden isn't mine to bear alone. It comes with the territory of being second-generation, embedded in skin and bones and legacies. My ancestors were born to nations thousands and thousands of miles away. I spring from a line of immigrants, one among countless dynasties who look or sound or exist differently from the rest. Our bodies unapologetically take up space among the throngs of people who blame us for invading their domains, as if the places they live are theirs and theirs alone. In the same way scientists have delegated kudzu, and other invasive plants and animals, as environmentally "other," so have white populations long diminished Asian, Black, Latino, and Indigenous bodies nationwide. We have been categorized, classified, and shrunken to labels that imply we are less than. We have been caught in a wave of prejudice that continues to swell with each passing day. It is fed by unease, for history shows us it is far more common to fear the unknown than to seek to understand it. We are free, but we are not. Like kudzu, we have been stamped as something to shun. Like kudzu, we have thrived in areas we are unwanted, despite resistance to our existence. Like kudzu, we have persisted.

A little over three years ago, the "vine that devoured the South" was nothing more than a passing fixture in the landscapes I knew so intimately, but also did not know at all. At once a stranger and a threat. Equally mystifying was the way it was completely ignored by the people around me, and yet sought out as target practice by those who dismissed it. Permaculturist Justin Holt finds that dissonance to be unique to kudzu. "It's a marginalized plant, a plant that loves the edges. And it is 'The Outsider,' or 'The Wild Thing,'" he said. Cofounder of Kudzu Culture, the modern-day reincarnation of the weed's long-defunct fan club helmed by Channing Cope, Holt finds nothing more compelling than the rich influence of kudzu embedded within the pages of the past, ecological scripture that pays homage to the impassioned people fighting to preserve its impact and the ways they work to rewrite its legacy. From one Indigenous community leader teaching neighbors to use it for paper to a husband-wife duo making a living by cooking with the vine's blossoms, Holt is among a handful of Americans who cherish the weed—despite what others may say about it. "We project our discomfort with not being really from here and having a play-space culture where we're really part of our ecosystem. We project that anxiety onto other plants and say, 'This plant belongs and this plant doesn't,'" he said.

Like a nebula of stars connecting a magnificent constellation in the infinite night sky, he belongs to an invisible nexus spanning the South, people who have been entwined together by a collective love for one emblematic perennial. An accidental community of sorts, one that remains outspoken and steadfast in the belief that kudzu, among other invasive plants, isn't the issue— we are. "There's a war on invasive species . . . but we are the invaders ourselves," declared Holt. "What we're trying to do is to basically spread the word that kudzu is not just this evil invader that we need to wage war against." Propelled by a white-hot rush of conviction, a near-constant state of being when he talks about the subject, the ecologist insists that the overabundance of kudzu in the US isn't the problem; the response to the overgrowth is. "The story of our somewhat schizophrenic relationship with kudzu comes down to a deeper thing, which is that, in the United States, we have a crisis of belonging," he added. The way he sees it, kudzu has been unfairly cast as a crippling force, when the real antagonists are the humans trying to eradicate it, punishing the very thing that defies all odds in its ability to survive. "By spending time looking at the world through the lens of kudzu, we can see that if there's an invasive species, it's the modern, industrial, culturally capitalist human." In his mind, kudzu is an incredible gift and a pristine resource, and it should be treated as such.

During that conversation with Holt something was said that burrowed its way into my mind, lingering like an incandescent star in the night sky, words that remained imprinted long after we said our goodbyes. It was a phrase that packed an altering punch, one that rattled my ribcage with the promise of something more. This string of purposeful words was what prompted me to dig further into one plant's thorny past and reevaluate the characterizations and societal perceptions that have taken a position of power concerning it. "Now it's, 'The vine that ate the South.' But if you look at what people were saying [about kudzu] in the 30s, and the 40s, they called it the 'Savior of the South,'" he said. That dichotomy echoes the layered complexities of America's shifting relationship with people like my parents, people like me. Where those who don't look like the rest are viewed with the same derision shown to an "alien invader." Welcomed, but unwanted. Celebrated, but condemned. The differences blur between a plant at the mercy of century-old rhetoric and the communities that remain the target of animosity. Like the sheets of kudzu that cover the topsoil around me, I didn't choose to be here. Like kudzu, I was brought here, by people who didn't know any better. So why must I hide my heritage, forced to the fringes of the place that is now my home? Kudzu is no more a harbinger of devastation, a damning curse that befell the land, than I am.

There is danger in humanizing plants that threaten other species to a point of no return. In no way should we repeat the mistakes we've made in the past—a bountiful weed like kudzu should never be planted with mass abandon, nor should it be presented as a sensational solution to depleted terrain. The evidence on its impact that does exist is unrelenting—when introduced to a new landscape, kudzu's inclination to overtake existing flora disrupts the biodiversity of an ecosystem. Biodiversity is indispensable to plant pollination, clean air, and water. According to the IPBES, an intergovernmental organization that provides policymakers worldwide with scientific assessments on biodiversity and ecosystem services, invasive species are among the top contributors to global biodiversity loss. When considering those implications, kudzu abruptly becomes antiheroic. In more ways than not, it is kryptonite to a healthy, thriving ecosystem. But it's also just a plant. It's not the diabolical, scheming antagonist we have made it out to be. And it's here in the South because of our ancestors, who brought it here, and our former leaders, who encouraged its expansion and created policies that allowed for its reign of abundance. The problem with kudzu in the US has more to do with how we define the vine than anything else. It is less about how we react to its presence, and more to do with how we misappropriate it. Americans have a track record of heaping unjust perceptions upon things that cannot defend themselves— be it plants or people. Racialized context and incriminating stereotypes have been intertwined into cultural embodiments of the vine. And that past is just as important to kudzu's narrative today as it is tomorrow.

As we move forward, we can change that. Perhaps these modifications could be simple and understated at first. Perhaps we can look to alter the way we speak about and decide to prioritize plants like kudzu—flora and fauna brought to the US, with origins in distant places. Perhaps we can try to invest further into better understanding things we don't. Above all, perhaps we can help elevate the underserved communities at the center of the stories we tell others, and ourselves. Kudzu is far more, culturally and climatically, than an overlooked weed that springs from southern ground, searching for other plants to overtake, like a fox hunting for prey. In many ways, it is an emblem of immigration, a flag of distinction, of otherness. It is a symbol of the racial and ethnic diversity that makes up the South, a reminder of its Eastern ancestry. To a few, it represents the opposite. To some, it *is* the South. It is at once everything and nothing at all, a crisis of belonging and a bridge to forgotten history. With every passing day, its meaning and message change, morphing to appease the eye of the beholder, the person gazing up at its blanket of imposing leaves cascading from the sky. To many, kudzu is magic. Although its past may be marred by shifting allegiance and societal dissonance, its path ahead is not without hope.

There are those who shirk the condemned legacy cast upon the vine. Those who fight hard to break the mold and the stereotype decided for them based on their surroundings. Those who not only look to the past to avoid making the same, tired mistakes, but who work tirelessly to help the communities around them do the same. You've met a handful of them in this book. Sometimes, that's all it takes. For as long as a few remember the origin behind the monstrous vine that gobbled up an entire region, as long as someone, somewhere whistles a tune of redemption into the listening wind, and as long as one child opens their heart to the treasure within a swath of striking greenery gleaming under the kiss of a sunburst, that will be enough. Those bristly vines will continue to thrive, wrapping acres of abandoned land in their embrace, shoots reaching across time and space. And, if we're lucky, our knowledge of what they do and why they do it will expand right alongside them. Whether it ends up covering millions of acres across the South, or no more than hundreds of thousands, kudzu will remain. It is a luminous, living reminder of the beauty that somehow always finds a way to thrive in the aftermath of our mistakes. It is a testament to tenacity. It is a nod to the very worst and the very best of humanity. Of every one of us.

Maybe that means there's still a chance—an opportunity for tides of change to wash over the communities that kudzu builds its ascending, green palaces around. Maybe one day the vines will represent the ways a society looked to its past, closed centuries of chapters based in bigotry, and started anew. Maybe they'll push people who once condemned them to open a blank, unwritten book and begin to tell a different tale. A story about a plant now at odds with a place that once sought it out. Where its presence brought communities together after forcing them apart. Where a nexus of extraordinary people found a blessing where others saw a curse. And a few use it to fight for a more just, kinder, more equal world.

Just one tendril, leaping toward the dawn of a better horizon, at a time.

A Note from the Author

We don't choose when we go from making memories to becoming one. When these hearts stop beating, blood no longer pumps through the veins that weave through us, and our brains begin to wind down to nothingness. All we can hope is to have a say in the little bit of the world we leave behind.

And that world is dying. As I write this, extreme heat waves, droughts, floods, and storms are sweeping over the 510 million kilometers that make up Earth's vast and spectacular surface. The consequences of human-induced climate change are manifesting in nearly infinite, almost uncountable, but very noticeable ways. Flames flickered at unbelievable heights in Siberia in recent years, a formerly frozen landscape ravaged by a blaze larger than all of the planet's other wildfires combined. Frequent droughts in the Horn of Africa are exacerbating a starving region's food crisis. Across the Global South, millions of people have been displaced by conflict triggered by climate shocks. From the United States to Australia, everyone, everywhere is facing the effects of our rapidly warming world. And those burdens are not borne equally.

To some, the realities of worsening climate change may still feel very far away. For many more, the opposite is true. I'm reminded of it every time I step outside, surveying the concrete landscape unfolding in front of me, ghosts of the lush greenery that once covered these roads vanishing in the throes of the latest heat wave. I'm reminded when it rains and cars carefully navigate Florida's heavily flooded streets, infrastructure crumbling as seas continue to rise, seeing firsthand what it's like when the people you love lose everything to a wall of water. I'm reminded when I report on the skyrocketing global number of climate refugees fleeing for their lives, pouring out of nations and neighborhoods with nowhere to go. I'm reminded as I witness our collective empathy for these kinds of human plights disappearing at the same horrifying, irreversible pace as one crucial ice sheet melting in Greenland.

Irreversible, these effects are expanding, but we still get to decide what the extent of the damage will look like. We are not passive bystanders in a waiting game, idly standing by as the world floods, melts, and goes up in flames around us. We can no longer act for prevention's sake, but we *can* act in the name of adaptation. We may have missed our shot to curtail climate change completely, but we still can decelerate the pace of global warming, and more equitably prepare for its indelible impacts. There is still a window of opportunity to dramatically curb human emissions of heat-trapping gases and black carbon, slowing down the rate our climate is changing and allowing us to forge a better future.

If this was a show, it would play out across all seven continents, one starring a cast of characters that spans more than seven billion. Ours is a dance without a certain ending, one that warns with each twist and turn that the choices we make today will set the stage for future generations to come. This will end either in applause, as humankind sacrifices a dangerous manner of living in order to strive on, or to the silence of an empty room, with nothing but lingering spirits serving as a veiled reminder of what, together, we willfully gave up. Our planet is degrading around us, but it doesn't have to be. Every moment ahead, every decision we make is a chance to get it right.

We owe it to the living world, as well as to ourselves, to stop and pay attention to the havoc each one of us is responsible for. Humans carve out lives where animals, plants, organisms, and the living, breathing cosmos around us are no more than objects—operating underneath the misguided belief that we are the only beings worthy of a habitat. If we move forward at our current pace of destruction, we will not only eviscerate other species' hopes of a better tomorrow, but our own. Without the preservation and protection of the thriving forests, bodies of water, and diverse ecosystems we so deeply rely on, we will miss our own shot at existence. We may not fully understand an entity that isn't man-made, but much like faith, we also don't need to comprehend the incomprehensible to worship it. We should not simply fight for the preservation of the earth, as if it were a thing. We are the earth. Every last part of every last one of us is tied to the planet we should all work to protect.

Without the natural world we rely on, the soils we stand on, the trees we look up to, and the oceans we gaze upon, we are lost. May we never forget that.

Acknowledgments

No one achieves anything alone. This book wouldn't be possible without the many, many wonderful people who helped shape it, who inspired it, who trusted me with their stories for it. This book also wouldn't be here without those who bolstered me through writing it.

Nothing but gratitude to the inimitable Jenny Keegan, LSU Press trade editor extraordinaire, who first came across my reporting on the artists, architects, and chefs working to embrace a quintessentially southern vine. Jenny, when you reached out to ask if I would ever consider writing a book about kudzu, I have never felt more seen or understood. Thank you for believing in a first-time author with a penchant for exclamation points. And my everlasting appreciation to Lyndsey Gilpin, managing editor of *Southerly*, who took the chance on my impassioned pitch—okay, *plea*—to compose that very story, to begin with, and then used her editorial wizardry to help me produce a piece that would change the trajectory of my professional career, and in many ways, my life.

All my thanks to Alisa Plant, freelance editor Dabian Witherspoon, James Wilson, Sunny Rosen, Michelle Neustrom, and everyone at LSU Press who devoted their time and talent to helping polish and promote this book into something I couldn't be more thrilled about. An extra note of gratitude to Catherine Kadair—thank you for your first-class edits and devotion to detail that helped bring out the very best in this.

My appreciation to all of the talented writers, researchers, and plant enthusiasts who have ever published anything having to do with kudzu. Your tireless devotion to the "vine that devoured the South" helped inform this endeavor of mine. To Saeed Jones, in particular, for writing a poem that still gives me goosebumps each time I re-read it. Your take on kudzu embodies so much of what I hope this book reflects. To Edward Francisco, for allowing me the honor of excerpting your powerful poetry. To Hongda Zheng, who so graciously helped translate and track down answers to my slew of queries about one mysterious weed in Chinese and Japanese literature. To Han Chen, who kindly aided in reviewing some of those references. To Cynthia Barnett, who answered all my burning questions about debut book-writing with compassion and encouragement. And all the gratitude to Emily Krieger, who fact-checked the hell out of this thing. Without you, "Gengis Con" might still live on.

I've been privileged to have worked with some of the most generous editors and mentors in the news industry over the course of my career. My thanks first and foremost to John Upton, who introduced me to climate coverage and taught me the importance of finding and telling the human stories at the heart of the science. I'm so grateful you took a chance on a new reporter with a fledgling portfolio. To David Nather, who elevated my reporting and writing to the next level. Under your tutelage, I was able to develop an entirely new beat covering the intersection of climate change and food security. I credit much of that to your editorial guidance. To Niala Boodhoo—not only are you a wildly successful journalist and newsroom leader, but you help elevate those around you. I can't thank you enough for all you do for women of color in this space. My appreciation to the talented humans behind the Society of Environmental Journalists, the Uproot Project, the Metcalf Institute, the Woods Hole Oceanographic Institution, and Oregon State University who facilitate some of the most revered fellowships in the environmental journalism game. I remain honored to have been included among your various 2022–2023 cohorts.

Finally, my profound gratitude to the immensely talented editors and reporters I've had the privilege of working with over the years. I appreciate each and every one of you.

How could I not jump at the chance to shout from the rooftops about those precious humans who fortified my support system during this process? I am profoundly grateful to Becky, who championed this story—and my ability to write it—from the very first day a burgeoning book deal began to take shape. Thank you for sparing no effort to travel with me to Ireland's fabled Cliffs of Moher that one time I needed a friend the most. I consider myself so lucky to know you, and I firmly believe we were sisters in another life. To Kaytlin, for being a support system from afar during the long, arduous, and fickle process that is writing a book, and the decade (!) of heartaches, successes, setbacks and moments of resilience we shared together beforehand. You are my oldest friend, and among the people I most cherish. To Sophie, Fiona and Sam, for keeping our WhatsApp banter—and kinship—thriving through international moves. To everyone else I didn't name who has influenced my trajectory in some way or another—thank you for inspiring me to have the audacity to do this. And last, but certainly not least, to Myra, Steve and Danny—for patiently listening to me talk about this project-in-progress and for consistently and enthusiastically asking for updates throughout.

Of course, endless gratitude to my family. My appreciation to my many, many relatives spread across many, many thousands of miles —*makasih*, *hvala*, *tack*, and *dankje*. In particular, my siblings—Raeza, Linaeya, Yriana, and Avel— roughly two-thirds of whom will likely never read this and won't even try to pretend they have. (I already forgive you.) How can I possibly figure out the right words to thank people who haven't just been my blood relations, but also my closest confidantes? Thanks for keeping me sane, grounded and humble. Sharing a dose of extra appreciation to the One With the Longest Name, for helping proofread a book proposal that allowed an inkling of a dream to become a reality, and to the One Who was Born Last, for the enthusiasm you brought to this inconceivable venture.

My deepest gratitude to my mom, for always believing in me and reminding me of my worth. In all the ways that matter, you have taught me the true meaning of strength and success. Without a doubt, you are the best person I know. All the thanks to my dad, for insisting I analyze those college-level essays in the sixth grade—a career as a journalist was pretty much a given after that. On a more serious note, thank you for teaching me my culture and how to tap into the unlimited power of ancestry. Your love for words is reflected in my own. And to both of my parents, for raising me in a way that honored your unique alchemy of heritages, beliefs, and native languages. Thank you for your stories of survival in the face of ignorance and adversity, and for teaching me to celebrate, not shy away from, the ways I stand apart.

There's no one who quite deserves a wholehearted "thank you" more than Alex. My "unofficial" editor; my first and my favorite beta reader; and my loudest cheerleader through all the rounds of reporting, drafting, editing, interviewing, and fact-checking as well as the sleepless nights plagued by self-doubt. If there's anyone who really understands what this experience was like, it's you. In a multitude of ways, through gestures big and small, you brought this project to fruition for me. My story with the vine that stole my heart starts with you, a lovely drive down a run-of-the-mill road blanketed in enigmatic tree canopies, and one swath of unforgettable kudzu. You are woven into every last piece of this, every last word on every last page. I could not have done it alone, and I will be plying you with café con leche in gratitude for a lifetime. *Aku cinta kamu.*

Full disclosure: I wrote this book without really knowing what I was getting into. I'm glad (if not more incredulous) that I managed to pull it together. Through it all, I couldn't stop wondering incessantly if I was the right person to do this. Sure, I'm a climate reporter with a penchant for plant ecology— but what made *me* the expert on all things kudzu? A few years down the line, I won't pretend that my imposter syndrome has dissipated or even that I am the most qualified to dig into the cultural and historical nuances behind the "vine that devoured the South." That said, I put a hell of a lot of work into filtering fact from fiction regarding an emblem so many love to hate. I moved heaven and earth to interview people dispersed across the country who have found their personal and professional lives changed by kudzu. And to those very people—who entertained my calls, emails, and questions; who carved hours out of their days to meet with me, to open up and be vulnerable; who trusted that I would communicate their life's work and experiences with the respect, empathy, and detail they all dearly deserve—thank you.

This is for you.

Notes

INTRODUCTION

2   **Set amidst green palaces:** James E. M. Watson, Oscar Venter, Jasmine R. Lee, Kendall R. Jones, John Robinson, Hugh P. Possingham, and James Allan, "Protect the Last of the Wild," *Nature* 563 (2018): 27–30, https://doi.org/10.1038/d41586-018-07183-6.

4   **A second-generation American:** "Second-Generation Americans: A Portrait of the Adult Children of Immigrants," Pew Research Center, last modified February 7, 2013, https://www.pewresearch.org/social-trends/2013/02/07/second-generation-americans/.

    producing racially ambiguous results: Alexandra Kleeman, "The Secret Toll of Racial Ambiguity," *New York Times* (November 22, 2021). https://www.nytimes.com/2021/10/20/magazine/rebecca-hall-passing.html.

    The older I got, the more: "Episcia Cupreata (Flame Violet)," North Carolina Extension Gardener Plant Toolbox, n.d., https://plants.ces.ncsu.edu/plants/episcia-cupreata/.

    Perhaps, in reality, nothing had changed: Joel Rose, "Talk of 'Invasion' Moves from the Fringe to the Mainstream of GOP Immigration Message," *NPR* (August 3, 2022), https://www.npr.org/2022/08/03/1115175247/talk-of-invasion-moves-from-the-fringe-to-the-mainstream-of-gop-immigration-mess.

I. A GRAND DEBUT

8   **"I feel like I'm truly grounded in the earth":** Beth Phillips, interview with author, February 15, 2022, Rogersville, AL.

    Kudzu was in Philadelphia: "Research Guides: Philadelphia's World Fair: Topics in Chronicling America: Introduction," Library of Congress Research Guides, n.d., https://guides.loc.gov/chronicling-america-worlds-fair-philadelphia.

    which cost more than $11 million: "Philadelphia Centennial Exposition | Trade Fair, Philadelphia, Pennsylvania, United States," *Encyclopedia Britannica* (July 20, 1998); "Philadelphia 1876," United States Centennial International Exhibition, National Gallery of Art, n.d., https://www.nga.gov/research/library/imagecollections/photographs-of-international-expositions/philadelphia-1876.html.

9   **A little over six years:** "Centennial Exhibition (1876)," *Encyclopedia of Greater Philadelphia* (February), https://philadelphiaencyclopedia.org/essays/centennial/.

Upwards of nine million people: Sandy Hingston, "10 Things You Might Not Know About the 1876 Centennial Exhibition," *Philadelphia Magazine* (May 10, 2016), https://www.philly mag.com/news/2016/05/10/centennial-exhibition-history/.

Also known as: "The Centennial Exhibition, Philadelphia, 1876," The Library Company of Philadelphia, n.d., https://www.lcpimages.org/centennial/.

Representatives from fifty-six: Sandy Hingston, "10 Things You Might Not Know About the 1876 Centennial Exhibition," *Philadelphia Magazine* (May 10, 2016), https://www.philly mag.com/news/2016/05/10/centennial-exhibition-history/.

Gently curved beams on a: "Digital Collections: Japanese Dwelling," Free Library of Philadelphia, n.d., https://libwww.freelibrary.org/digital/item/2146.

In control at the time: Steven Ujifusa, "Japan-a-Mania at the Centennial," *PhillyHistory Blog* (May 12, 2010), https://blog.phillyhistory.org/index.php/2010/05/japan-a-mania-at-the -centennial/.

10   **Notable attendees raved about:** Russell Frank Weigley, Nicholas B. Wainwright, and Edwin Wolf, *Philadelphia: A 300 Year History* (W. W. Norton & Company, 1982).

2. PORTAL INTO THE PAST

16   **"Gosh, it's—I'm such—it can go so many different ways":** Lauren Bacchus, interview with author, February 2, 2022, Asheville, NC.

17   **Formally introduced to the US:** William Shurtleff, *The Book of Kudzu: A Culinary & Healing Guide* (Avery, 1985).

Then, in 1884 and 1885: Anonymous, "The Smithsonian Institution in the 1884 New Orleans World's Fair," Smithsonian Institution Archives, n.d., https://siarchives.si.edu/blog /smithsonian-insitution-1884-new-orleans-world%E2%80%99s-fair; Anna E. Eskridge and Derek H. Alderman, "Alien Invaders, Plant Thugs, and the Southern Curse: Framing Kudzu as Environmental Other through Discourses of Fear," *Southeastern Geographer* 50, no. 1 (n.d.): 110–29, https://doi.org/10.1353/sgo.0.0073.

For those in attendance: Anna E. Eskridge and Derek H. Alderman, "Alien Invaders, Plant Thugs, and the Southern Curse: Framing Kudzu as Environmental Other through Discourses of Fear," *Southeastern Geographer* 50, no. 1 (2010): 110–29, https://doi.org/10.1353 /sgo.0.0073.

Behind the scenes: Anna E. Eskridge and Derek H. Alderman, "Alien Invaders, Plant Thugs, and the Southern Curse: Framing Kudzu as Environmental Other through Discourses of Fear," *Southeastern Geographer* 50, no. 1 (2010): 110–29, https://doi.org/10.1353/sgo.0.0073.

18   **Its seeds were the first element:** "History and Use of Kudzu in the Southeastern United States," Alabama Cooperative Extension System, March 8, 2022, https://www.aces.edu/blog/topics /forestry-wildlife/the-history-and-use-of-kudzu-in-the-southeastern-united-states/.

Around 1902, one influential: Robert J. Hill, "Kudzu-Vine, Pueraria Lobata (Willd.) Ohwi."

Regulatory Horticulture 11, no. 1 (1985), https://www.agriculture.pa.gov/Plants_Land_Water/PlantIndustry/NIPPP/Documents/kudzu%20article.pdf.

the origin of the US Department: Kristen Hinman, "Kudzu—Japan's Wonder Vine," HistoryNet, September 29, 2017, https://www.historynet.com/kudzu-japans-wonder-vine/.

19   **According to his written accounts:** "Full Text of 'David Fairchild (1938) The World Was My Garden'," n.d., https://archive.org/stream/david-fairchild-the-world-was-my-garden/David%20Fairchild%20The%20World%20Was%20My%20Garden_djvu.txt.

In 1993, the federal government: "A Brief History of NRCS," Natural Resources Conservation Service, n.d., https://www.nrcs.usda.gov/about/history/brief-history-nrcs.

Farming was the occupation listed for almost one-third: "Farm Population Lowest Since 1850's," *New York Times* (July 20, 1988), https://www.nytimes.com/1988/07/20/us/farm-population-lowest-since-1850-s.html.

Historical records date back to 1909: Chris Lawrence and Maurice G. Cook, 2015. "Brief Biography of Hugh Hammond Bennett," National Resources Conservation Service/USDA, August 13, 2015, https://efotg.sc.egov.usda.gov/references/public/VA/ER1a_H_H_Bennett_Biography_2019B.pdf.

Brian Barth at *Modern Farmer:* Brian Barth, "How Kudzu, 'The Vine That Ate the South,' Put Southern Agriculture on the Skids," Modern Farmer, October 19, 2018, https://modernfarmer.com/2016/10/kudzu/.

The vine's leaves and root system: Jacob Johanson, "Invasive Species," Longwood, November 10, 2016, http://blogs.longwood.edu/webbbd/.

"enhance soil fertility": Dylan Glass and Safaa Al-Hamdani, "Kudzu Forage Quality Evaluation as an Animal Feed Source," *American Journal of Plant Sciences* 7, no. 4 (2016): 702–7, https://doi.org/10.4236/ajps.2016.74063.

20   **That was all the federal government:** Richard Hornbeck, "The Enduring Impact of the American Dust Bowl: Short- and Long-Run Adjustments to Environmental Catastrophe," *American Economic Review* 102, no. 4 (2012), https://www.jstor.org/stable/23245462.

The first major attempt: Richard J. Blaustein, "Kudzu's Invasion into Southern United States Life and Culture," US Forest Service Research and Development, 2001, https://www.fs.usda.gov/research/treesearch/6184.

The year 1935 marked the first widespread efforts: Bill Finch, "The True Story of Kudzu, the Vine That Never Truly Ate the South," *Smithsonian Magazine* (August 24, 2015), https://www.smithsonianmag.com/science-nature/true-story-kudzu-vine-ate-south-180956325/.

In the 1930s and 1940s: Derek H. Alderman and Donna D. Alderman, "Kudzu: A Tale of Two Vines," *Southern Cultures* 7, no. 3 (2001): 49–64, https://doi.org/10.1353/scu.2001.0030.

Accounts differ on: "History and Use of Kudzu in the Southeastern United States," Alabama Cooperative Extension System, March 8, 2022, https://www.aces.edu/blog/topics/forestry-wildlife/the-history-and-use-of-kudzu-in-the-southeastern-united-states/.

The US government offered incentives: "Inflation Rate between 1940–2023," CPI Inflation Calculator, n.d., https://www.in2013dollars.com/us/inflation/1940?amount=8#:~:text=%248%20in%201940%20is%20equivalent,cumulative%20price%20increase%20of%202%2C019.98%25.

21   **The agency was joined by:** "The Civilian Conservation Corps," US National Park Service, n.d., https://www.nps.gov/articles/the-civilian-conservation-corps.htm.

Farmer, journalist, and radio show host: Derek H. Alderman, "Channing Cope and the Making of a Miracle Vine," *Geographical Review* (April 2004), https://www.jstor.org/stable /30033969.

In the lead-up: Ryan Engelman, "The Second Industrial Revolution, 1870–1914," *US History Scene* (October 25, 2020), https://ushistoryscene.com/article/second-industrial-revolution/.

City populations nationwide: "U.S. History Primary Source Timeline," Library of Congress, n.d., https://www.loc.gov/classroom-materials/united-states-history-primary-source-time line/rise-of-industrial-america-1876-1900/overview/.

22   **Wrapping its sheets:** Richard J. Blaustein, "Kudzu's Invasion into Southern United States Life and Culture," US Forest Service Research and Development, 2001, https://www.fs.usda .gov/research/treesearch/6184.

kudzu had even crept over railroad tracks: Mark Allen Stewart, "Cultivating Kudzu: The Soil Conservation Service and the Kudzu Distribution Program," *Georgia Historical Quarterly* 81, no. 1 (1997), https://www.jstor.org/stable/40583548.

The purportedly "uncontrollable" rate: "Climate Change Indicators: U.S. and Global Temperature," US EPA, August 1, 2022, https://www.epa.gov/climate-indicators/climate -change-indicators-us-and-global-temperature.

The federal government responded to: Mark Stewart, "Kudzu." *New Georgia Encyclopedia* (last modified Aug 26, 2019), https://www.georgiaencyclopedia.org/articles/geography -environment/kudzu/.

displaying the same urgency: Emilie Reas, "Small Animals Live in a Slow-Motion World," *Scientific American Mind* (July 1, 2014), https://doi.org/10.1038/scientificamericanmind071 4-11a.

It initially removed the noxious species: John W. Everest, James H. Miller, Donald M. Ball, and Mike Patterson, "Kudzu in Alabama: History, Uses and Control," Alabama Cooperative Extension System, 1999, https://www.srs.fs.usda.gov/pubs/ja/ja_everest001.pdf.

It would take another: D.K. Jewett, "Characterizing Specimens of Kudzu and Related Taxa with RAPD's," US Forest Service Research and Development, 2003, https://www.fs.usda .gov/research/treesearch/6039.

To some degree: Kathleen S. Lowney and Joel Best, "Floral Entrepreneurs: Kudzu as Agricultural Solution and Ecological Problem," *Sociological Spectrum* 18, no. 1 (1998): 93–114, https://doi.org/10.1080/02732173.1998.9982186.

23   **It's a plant that can be traced:** "葛の特性と歴史・文化 ｜ 廣久葛本舗," Kyusuke, n.d., http://kyusuke.co.jp/history/index.html.

A 1973 journal article by Chinese botanist Hsuan Keng: Hsuan Keng, "Economic Plants of Ancient North China as Mentioned in 'Shih Ching' (Book of Poetry)," *Economic Botany* 28, no. 4 (1974), https://www.jstor.org/stable/4253534.

The oldest existing collection of Chinese poetry: "Shih Ching—The Poetry Classic," *Poetry of China* (n.d.), http://poetrychina.net/poets/shih_ching.

In *The Book of Kudzu*: William Shurtleff, *The Book of Kudzu: A Culinary & Healing Guide* (Avery, 1985).

likely first recorded origin of kudzu in Japan: "Man'yō-Shū | Japanese Anthology," *Encyclopaedia Britannica* (July 20, 1998).

Translated as "Collection of Ten Thousand Leaves": Mark Cartwright, "Manyoshu," *World History Encyclopedia* (May 2023), https://www.worldhistory.org/Manyoshu/.

24   On Blossom: Temca, "Tag: Kudzu." Waka Poetry, n.d, https://www.wakapoetry.net/tag /kudzu/.

25   Just outside of downtown Asheville: Barbara Durr, "Black Home Ownership and the Promise of Reparations," *Asheville Citizen Times* (March 2, 2021), https://eu.citizen-times.com /story/news/local/2021/03/02/asheville-watchdog-black-home-ownership-and-promise -reparations/6880327002/.

26   Someday, not too far: E. Ann Carson, Danielle H. Sandler, Renuka Bhaskar, Leticia E. Fernandez, and Sonya R. Porter, "Employment of Persons Released from Federal Prison in 2010," Bureau of Justice Statistics, Office of Justice Programs, December 2021, https://bjs .ojp.gov/content/pub/pdf/eprfp10.pdf.

27   she remembers drinking kudzu: WAWAZA, "How to Make Kuzu-Yu (Arrowroot Tea) the Japanese Way," *WAWAZA* (n.d.), https://wawaza.com/pages/how-to-make-kuzu-yu-arrow root-tea-the-japanese-way/.

3. KUDZU VERSUS THE SOUTH

29   Like a migrating leatherback turtle: NOAA Fisheries, "Leatherback Turtle," NOAA, August 10, 2020, https://www.fisheries.noaa.gov/species/leatherback-turtle.

A large, often trifoliate-leaved: "History and Use of Kudzu in the Southeastern United States," Alabama Cooperative Extension System, March 8, 2022, https://www.aces.edu/blog/topics /forestry-wildlife/the-history-and-use-of-kudzu-in-the-southeastern-united-states/.

A member of the *Fabaceae*: Ka Sing Wong, George Q. Li, Kong M. Li, Valentina Razmovski-Naumovski, and Kelvin K. W. Chan, "Kudzu Root: Traditional Uses and Potential Medicinal Benefits in Diabetes and Cardiovascular Diseases," *Journal of Ethnopharmacology* 134, no. 3 (2011): 584–607. https://doi.org/10.1016/j.jep.2011.02.001.

30   the Swiss botanist: "Pueraria Montana (Kudzu)," North Carolina State Extension, n.d., https://plants.ces.ncsu.edu/plants/pueraria-montana/.

native to countries in East and South Asia: Thomas Forney and Glenn Miller, "Oregon Department of Agriculture Plant Pest Risk Assessment for Kudzu, Pueraria Montana," Oregon Department of Agriculture, 2013, https://www.oregon.gov/ODA/shared/Documents /Publications/Weeds/PlantPestRiskAssessmentKudzu.pdf.

including China, Japan, Korea, Thailand: Steve Csurhes and State of Queensland, "Kudzu Risk Assessment," Queensland Government, 2016, https://www.daf.qld.gov.au/__data/assets /pdf_file/0004/74137/IPA-Kudzu-Risk-Assessment.pdf.

Kudzu is known as: Shurtleff, William, *The Book of Kudzu: A Culinary & Healing Guide* (Avery, 1985).

At first glance: "Smilax Rotundifolia," Forestry Service, USDA, n.d., https://www.fs.usda.gov/database/feis/plants/vine/smirot/all.html.

Upon closer inspection: "Poison Ivy," NYS Dept. of Environmental Conservation, n.d., https://www.dec.ny.gov/animals/105384.html#:~:text=Poison%20ivy%20climbs%20with%20black,and%20it%20does%20not%20twine.

These less-than-an-inch-long flowers: "Pueraria Montana (Kudzu)," North Carolina State Extension, n.d., https://plants.ces.ncsu.edu/plants/pueraria-montana/.

The buds imitate the rainbow: "Kudzu," New York Invasive Species (IS) Information, July 2, 2019, https://nyis.info/invasive_species/kudzu/.

The tap root is said to grow: "Center for Aquatic and Invasive Plants," University of Florida IFAS, n.d., https://plants.ifas.ufl.edu/plant-directory/pueraria-montana/.

Although reports vary: The Forest Health Technology Enterprise Team, "New Invaders of the Northwest," Bugwood Center for Invasive Species and Ecosystem Health, US Department of Agriculture, 2015, https://bugwoodcloud.org/resource/files/14874.pdf.

In a single day: "Kudzu," New York Invasive Species (IS) Information, July 2, 2019, https://nyis.info/invasive_species/kudzu/.

kudzu flourishes in disturbed and degraded areas: Richard J. Blaustein, "Kudzu's Invasion into Southern United States Life and Culture," US Forest Service Research and Development, 2001, https://www.fs.usda.gov/research/treesearch/6184.

Similar to marigolds: "Marigolds," Almanac.com. n.d., https://www.almanac.com/plant/marigolds#:~:text=Marigolds%20thrive%20in%20full%20sunshine,and%20won%27t%20bloom%20well.

or black-eyed Susan flowers: "How to Grow Black-Eyed Susans," MiracleGro, n.d., https://miraclegro.com/en-us/growing-plants/flowers/how-to-grow-black-eyed-susans.html#:~:text=Black%2Deyed%20Susans%20grow%20best,and%20spreading%20toward%20the%20light.

kudzu does best in full sunlight: Invasive Species Coordinator, "Kudzu," Missouri Department of Conservation, 2020, https://mdc.mo.gov/sites/default/files/2020-04/Kudzu.pdf.

31   **In 2021, she published a paper:** Gina Profetto and Jerome J. Howard, "Plant Community Responses to Kudzu (Pueraria Montana) Invasion in a Southern Upland Forest," *Journal of the Torrey Botanical Society* 149, no. 1 (2021), https://doi.org/10.3159/torrey-d-21-00019.1.

"I had no idea what the flora looked like in Louisiana": Gina Profetto, interview with author, June 16, 2022, Tifton, GA.

Study authors Irwin Forseth and Anne Innis wrote: Irwin N. Forseth and Anne F. Innis, "Kudzu (*Pueraria Montana*): History, Physiology, and Ecology Combine to Make a Major Ecosystem Threat," *Critical Reviews in Plant Sciences* 23, no. 5 (2004): 401–13, https://doi.org/10.1080/07352680490505150.

33   **What their research revealed:** Rachel Collins, Carolyn A. Copenheaver, Jacob N. Barney, and Philip J. Radtke, "Using Invasional Meltdown Theory to Understand Patterns of Invasive Richness and Abundance in Forests of the Northeastern USA," *Natural Areas Journal* 40, no. 4 (2020), https://doi.org/10.3375/043.040.0406.

Kudzu can survive almost anywhere: "Kudzu (Pueraria Montana)," Plants Database, National Gardening Association, n.d., https://garden.org/plants/view/206030/Kudzu-Pueraria-montana/.

Casting a substantial layer: "Kudzu," University of Florida Institute of Food and Agricultural Sciences, n.d., https://gardeningsolutions.ifas.ufl.edu/care/weeds-and-invasive-plants/kudzu.html.

34   **Compounding these impacts:** "History and Use of Kudzu in the Southeastern United States," Alabama Cooperative Extension System, March 8, 2022, https://www.aces.edu/blog/topics/forestry-wildlife/the-history-and-use-of-kudzu-in-the-southeastern-united-states/.

Asian soybean rust: Caspar Langenbach, Ruth Campe, Sebastian Beyer, André C. Müller, and Uwe Conrath,. "Fighting Asian Soybean Rust," *Frontiers in Plant Science* 7 (June 2016), https://doi.org/10.3389/fpls.2016.00797.

The code to its relentless survival: "History and Use of Kudzu in the Southeastern United States," Alabama Cooperative Extension System, March 8, 2022, https://www.aces.edu/blog/topics/forestry-wildlife/the-history-and-use-of-kudzu-in-the-southeastern-united-states/.

According to New York Invasive Species Information (NYIS.info): "Kudzu," New York Invasive Species (IS) Information, July 2, 2019, https://nyis.info/invasive_species/kudzu/.

The vine has been spotted: "USDA Plants Database," n.d., http://plants.usda.gov/.

Research suggests the vine: "Kudzu," Iowa Department of Natural Resources., n.d., https://www.iowadnr.gov/portals/idnr/uploads/forestry/kudzu.pdf?amp;tabid=1242.

Climate change is triggering: Ayurella Horn-Muller, "The Role of Kudzu in Architecture, Cuisine, and Culture," *Southerly* (January 14, 2022), https://southerlymag.org/2021/03/20/the-role-of-kudzu-in-architecture-cuisine-and-culture/.

35   **The mysteries at the heart:** USDA Agricultural Research Service, "Combined Control Tactics Remove Kudzu Faster," *AgResearch Magazine* 64, no. 7 (July 2016), https://agresearchmag.ars.usda.gov/2016/jul/kudzu/.

36   **In the South, especially in Georgia, Alabama, and Mississippi:** Richard J. Blaustein, "Kudzu's Invasion into Southern United States Life and Culture," US Forest Service Research and Development, 2001, https://www.fs.usda.gov/research/treesearch/6184.

Nowadays, kudzu is: Paulina Harron, Omkar Joshi, Christopher B. Edgar, Shishir Paudel, and Arjun S. Adhikari, "Predicting Kudzu (Pueraria Montana) Spread and Its Economic Impacts in Timber Industry: A Case Study from Oklahoma," *PLOS One* 15, no. 3 (2020): e0229835, https://doi.org/10.1371/journal.pone.0229835.

37   **"Oh, it's everywhere here":** Heather Coiner, interview with author, September 14, 2021, Roseland, VA.

38   **Undeterred, the researcher:** "Kudzu," Ministry of Natural Resources and Forestry, Ontario, Canada, last updated August 2, 2022, https://www.ontario.ca/page/kudzu#:%7E:text=The%20most%20extensive%20infestations%20have,Leamington%2C%20ontario%2C%20on%20the%20north.

What Coiner and her team unearthed: Heather A. Coiner, Katharine Hayhoe, Lewis H. Ziska, Jeff Van Dorn, and Rowan F. Sage, "Tolerance of Subzero Winter Cold in Kudzu (Pueraria Montana Var. Lobata)," *Oecologia* 187, no. 3 (2018): 839–49, https://doi.org/10.1007/s00442-018-4157-8.

Research published in 1969: F.T. Wolf, "Photosynthetic and Respiratory Rates as Influenced by Temperature and Light Intensity," *EurekaMag* (January 1969), https://eurekamag .com/research/014/604/014604253.php.

39   **She isn't the first to:** Robert J. Hill, "Kudzu-Vine, Pueraria Lobata (Willd.) Ohwi," *Regulatory Horticulture* 11, no. 1 (1985), https://www.agriculture.pa.gov/Plants_Land_Water/Plant Industry/NIPPP/Documents/kudzu%20article.pdf.

40   **"There's not this vast expanse of literature where people have evaluated":** Matthew Fry, interview with author, September 29. 2021, Ithaca, NY.

41   **The USDA Agricultural Research Service (ARS):** "Combined Control Tactics Remove Kudzu Faster," *AgResearch Magazine* 64, no. 7 (n.d.), https://agresearchmag.ars.usda.gov/2016/jul /kudzu/.

     although a 2016 ARS study: Mark A. Weaver, C. Douglas Boyette, and Robert E. Hoagland, "Rapid Kudzu Eradication and Switchgrass Establishment through Herbicide, Bioherbicide and Integrated Programmes," *Biocontrol Science and Technology* 26, no. 5 (2016): 640–50, https://doi.org/10.1080/09583157.2016.1141175.

     Herbicides are a widely deployed option: Sandra Avant, "A Faster Way to Get Rid of Kudzu," USDA Agricultural Research Service, July 13, 2016, https://www.ars.usda.gov/news-events /news/research-news/2016/a-faster-way-to-get-rid-of-kudzu/.

42   **reliance on the weed killer:** Vincent Fugère, Marie-Pier Hébert, Naíla Barbosa Da Costa, Charles C.Y. Xu, Rowan D. H. Barrett, Beatrix E. Beisner, Graham Bell, et al., "Community Rescue in Experimental Phytoplankton Communities Facing Severe Herbicide Pollution," *Nature Ecology and Evolution* 4, no. 4 (2020): 578–88, https://doi.org/10.1038/s41559 -020-1134-5.

     muted effectiveness on herbicide-resistant weeds: H. Claire Brown, "How Superweeds Like Palmer Amaranth Are Changing Agriculture," *New York Times* (August 18, 2021), https://www.nytimes.com/2021/08/18/magazine/superweeds-monsanto.html.

     A whitepaper published by the Virginia Department of Conservation and Recreation: "Invasive Alien Plant Species of Virginia," Virginia Department of Conservation and Recreation and Virginia Native Plant Society, n.d., https://www.dcr.virginia.gov/natural-heri tage/document/fspulo.pdf.

4. BEAUTY FROM THE BEAST

45   **"Someone looked out the [classroom] window":** Will May, interview with author, October 19, 2021, Winfield, AL.

     As they read up on the weed's background: Safaa H. Al-Hamdani and David Marc Ponder, "Evaluation and Comparison of Selected Antioxidant Concentrations in Kudzu and Three Common Food Sources," *Journal of the Alabama Academy of Science* 85, no. 1 (2014).

     a source of antioxidants: Eun-Jung Son, Jong Ho Yoon, Bong-Jeun An, Do Yun Lee, Ji-Min Cha, Gyeong-Yup Chi, and Dongseon Kim, "Comparison among Activities and Isoflavonoids from Pueraria Thunbergiana Aerial Parts and Root," *Molecules* 24, no. 5 (2019): 912, https://doi.org/10.3390/molecules24050912.

46   **Curd instructed the seventh graders:** "International Space Station Legal Framework," The European Space Agency, n.d., https://www.esa.int/Science_Exploration/Human_and _Robotic_Exploration/International_Space_Station/International_Space_Station_legal _framework.

      was one of a kind: Freda Curd, interview with author, October 19, 2021, Hackleburg, AL.

      May and his fellow "space kudzu pioneers": Will May, Seth Birdsong, Cole Kirkpatrick, Banks Roebuck, Izzy Stewart, and Freda Curd, "The Effects of Microgravity on the Germination of Kudzu Seeds," Winfield Middle School, 2017.

      In 1979, NASA scientists: "Plants Clean Air and Water for Indoor Environments," NASA Spinoff, n.d., https://spinoff.nasa.gov/Spinoff2007/ps_3.html; "NASA Technical Reports Server (NTRS) 19800009283: Energy from Aquatic Plant Wastewater Treatment Systems: NASA Technical Reports Server (NTRS): Free Download, Borrow, and Streaming," Internet Archive, September 1, 1979, https://archive.org/details/NASA_NTRS_Archive_19800 009283/page/n1/mode/2up.

48   **On April 27, 2011:** "Hackleburg Tornado—April 27, 2011," NOAA's National Weather Service, n.d., https://www.weather.gov/bmx/event_04272011hackleburg.

49   **Once the catastrophic:** Andrea Lindenberg, "Hackleburg Was Never the Same after the Tornado. A Decade Later, Residents Continue to Lean on One Another." CBS 42, April 27, 2021, https://www.cbs42.com/special-reports/hackleburg-residents-leaned-on-one-another -to-get-through-tornado-10-years-ago/.

      including some of the lush vegetation: "Census Profile: Hackleburg, AL," Census Reporter, n.d., https://censusreporter.org/profiles/16000US0132560-hackleburg-al/.

      A little over three hundred miles: Joan Vannorsdall, "Walhalla, South Carolina: Deep in Cherokee and German Pasts," *Blue Ridge Country* (April 2021). https://blueridgecountry.com /departments/our-blue-ridge-towns/walhalla-sc/.

      more than four thousand residents: "Explore Census Data," US Census Bureau., n.d., https://data.census.gov/profile?g=160XX00US4574095.

      "It greets people when they come in": Nancy Basket, interview with author, June 22, 2022, Walhalla, SC.

51   **All elements of nature:** "The Myths and Legends of the Cherokee People." Georgia Stories, PBS LearningMedia, 2022, https://florida.pbslearningmedia.org/resource/cba6b627-139f -4161-9ac4-60735bd3dabc/georgia-stories-cherokee-myths-and-legends/.

      Not just for themselves: David E. Wilkins, "How to Honor the Seven Generations," ICT, last updated September 12, 2018, https://ictnews.org/archive/how-to-honor-the-seven -generations.

      Trees are entities worthy: Kathy Van Buskirk, "Cherokee Myths, Legends and Superstitions," *New Statesman* (September 27, 2015), https://www.newstatesman.com/politics/2007/04/ cherokee-myths-culture-stories.

52   **Someone else once told her:** "Huaca | Inca Religion," *Encyclopedia Britannica* (July 20, 1998).

53   **For more than fifteen years:** "History and Use of Kudzu in the Southeastern United States," Alabama Cooperative Extension System, March 8, 2022, https://www.aces.edu/blog/topics /forestry-wildlife/the-history-and-use-of-kudzu-in-the-southeastern-united-states/.

"Kudzu is everywhere, and most people recognize it": Nancy Loewenstein, interview with author, August 27, 2021, Auburn, AL.

5. FOREST TO FORK

56   With the exception of its seeds and seed pods: Christopher Hassiotis, "Kudzu Is Edible. Why Aren't We Eating It?" *Atlanta Journal-Constitution* (August 1, 2018), https://www.ajc.com/entertainment/dining/kudzu-edible-why-aren-eating/BXAct9CtIshpWaB8f9D2PO/.
To this day: "History and Use of Kudzu in the Southeastern United States," Alabama Cooperative Extension System, March 8, 2022, https://www.aces.edu/blog/topics/forestry-wildlife/the-history-and-use-of-kudzu-in-the-southeastern-united-states/.
The blossoms tend to be: "Kudzu," New York Invasive Species (IS) Information, July 2, 2019, https://nyis.info/invasive_species/kudzu/.
The latter can account for: Gaixia Zhang, Jinxin Liu, Mei Hua Gao, Weijun Kong, Qing Zhao, Linchun Shi, and Qiuling Wang, "Tracing the Edible and Medicinal Plant Pueraria Montana and Its Products in the Marketplace Yields Subspecies Level Distinction Using DNA Barcoding and DNA Metabarcoding," *Frontiers in Pharmacology* 11 (March 2020), https://doi.org/10.3389/fphar.2020.00336.
In a 2009 *Kitchn* article: Kathyrn Hill, "Did You Know You Can Eat Kudzu?" *Kitchn* (August 11, 2009), https://www.thekitchn.com/did-you-know-you-can-eat-kudzu-92488.
From Georgia to Tennessee: Ligaya Mishan and Kyoko Hamada, "When Invasive Species Become the Meal," *New York Times* (October 3, 2020), https://www.nytimes.com/2020/10/02/t-magazine/eating-invasive-species.html.
Published in *Eat the Invaders:* "Kudzu," Eat the Invaders, January 19, 2012, http://eattheinvaders.org/kudzu/.
Another source of: Angela Gillaspie, "Southern Angel's Kudzu Recipes Garden Shed," last revised July 26, 2016, http://www.southernangel.com/food/kudzurcp.html.

57   Meanwhile, an assortment of recipes: "Kudzu Recipes." Food.com, Warner Bros. Discovery, Inc., n.d., https://www.food.com/search/kudzu.
The vine is also popular in: Lauren Panoff, "Kudzu Root: Benefits, Uses, and Side Effects," Healthline, February 8, 2021, https://www.healthline.com/nutrition/kudzu-root#what-it-is.
"I lived on this little street": Mimi Maumus, interview with author, March 8, 2022, Athens, GA.

58   In 2014, a report published in: Safaa H. Al-Hamdani and David Marc Ponder, "Evaluation and Comparison of Selected Antioxidant Concentrations in Kudzu and Three Common Food Sources," *Journal of the Alabama Academy of Science* 85, no. 1 (2014), https://go.gale.com/ps/i.do?id=GALE%7CA402738545&sid=googleScholar&v=2.1&it=r&linkaccess=abs&issn=00024112&p=AONE&sw=w&userGroupName=anon%7Ef37a42d7&aty=open+web+entry.

59   botanist for the Department of Agriculture: John Motyka, "James Duke, 88, Globe-Trotting Authority on Healing Plants, Is Dead," *New York Times* (December 5, 2018), https://www.nytimes.com/2018/12/05/obituaries/james-duke-dead.html.

Duke found that: "Pueraria Lobata (Willd.) Ohwi," Purdue University Department of Horticulture and Landscape Architecture, last updated January 8, 1998, https://www.hort.purdue.edu/newcrop/duke_energy/Pueraria_lobata.html.

61 **Restaurant and food-service sales:** Rachel King, "More than 110,000 Eating and Drinking Establishments Closed in 2020," *Fortune* (January 26, 2021), https://fortune.com/2021/01/26/restaurants-bars-closed-2020-jobs-lost-how-many-have-closed-us-covid-pandemic-stimulus-unemployment/.

It was there, in the "teensy little town": " Roopville town, Georgia," Explore Census Data, United States Census Bureau, 2021, https://data.census.gov/cedsci/profile?g=1600000US1366696.

6. COLLECTING AND CONNECTING

65 **"He didn't know that they were invasive, or care":** Joe Roman, interview with author, January 13, 2021, Richmond, VT.

66 **His thesis focused on:** "Details," Texas Invasive Species Institute, 2014, http://www.tsus-invasives.org/home/database/carcinus-maenas.

by hitching a ride on sailing ships: "Invasion Biology Introduced Species Summary Project—Columbia," Columbia University, n.d., http://www.columbia.edu/itc/cerc/danoff-burg/invasion_bio/inv_spp_summ/Carcinus_maenas.htm.

A 1998 study by: Andrew N. Cohen and James T. Carlton, "Accelerating Invasion Rate in a Highly Invaded Estuary," *Science* 279, no. 5350 (1998): 555–58, http://www.jstor.org/stable/2894042.

The UC Riverside Center: "The Delta," California Department of Water Resources, n.d., https://water.ca.gov/water-basics/the-delta.

Ranked number 18 out of the: Invasive Species Specialist Group, Species Survival Commission, and International Union for Conservation of Nature, "GISD," Global Invasive Species Database, 2023, http://www.iucngisd.org/gisd/100_worst.php.

Devouring other invertebrates: "What Are Marine Invasive Species?" California Department of Fish and Wildlife, May 30, 2023, https://wildlife.ca.gov/OSPR/Science/Marine-Invasive-Species-Program/Definition.

Just ten years after: Joe Roman, "Diluting the Founder Effect: Cryptic Invasions Expand a Marine Invader's Range," *Proceedings of the Royal Society B: Biological Sciences* 273, no. 1600: (2006): 2453–59. https://doi.org/10.1098/rspb.2006.3597.

67 **Not to be confused with:** Susan Pike, "Periwinkles Cute, but Invasive," *Seacoastonline* (July 22, 2016), https://eu.seacoastonline.com/story/lifestyle/2016/07/23/periwinkles-cute-but-invasive/27443874007/.

A 2009 study published: Susan H. Brawley, James A. Coyer, April M. H. Blakeslee, Galice Hoarau, Ladd E. Johnson, James E. Byers, Wytze T. Stam, and Jeanine L. Olsen, "Historical Invasions of the Intertidal Zone of Atlantic North America Associated with Distinctive Patterns of Trade and Emigration," *Proceedings of the National Academy of Sciences of the United States of America* 106, no. 20 (2009): 8239–44, https://doi.org/10.1073/pnas.0812300106.

In the Gulf of Maine: Sue Pike, "Three Kinds of Periwinkles Live in Gulf of Maine," *Portsmouth Herald* (July 19, 2016), https://www.seacoastonline.com/story/news/local/2016/07/19/three-kinds-periwinkles-live-in/27466916007/.

69   **Belonging to a large genus of flowering plants:** "Dandelion," Cornell College of Agriculture and Life Sciences, n.d., https://cals.cornell.edu/weed-science/weed-profiles/dandelion.

thought to be purposefully cultivated: Patty Wetli, "The Dandelion's Fall From Grace Has Been a Doozy. Can This Weed Become a Flower Again?" *WTTW News* (May 15, 2020), https://news.wttw.com/2020/05/14/dandelion-weed-flower-history.

Indigenous communities and European colonists both incorporated: Charlotte Bringle Clarke, "Edible and Useful Plants of California," University of California Press, February 1978, https://www.ucpress.edu/book/9780520032675/edible-and-useful-plants-of-california.

Unlike kudzu, ecologists: "Explore the Taxonomic Tree," US Fish and Wildlife Service, n.d., https://www.fws.gov/taxonomic-tree/37891.

70   **A 2017 study:** Colleen M. Synk, Brent F. Kim, Charles C. Davis, James Harding, Virginia Rogers, Patrick T. Hurley, Marla R. Emery, and Keeve E. Nachman, "Gathering Baltimore's Bounty: Characterizing Behaviors, Motivations, and Barriers of Foragers in an Urban Ecosystem," *Urban Forestry & Urban Greening* 28 (December 2017): 97–102, https://doi.org/10.1016/j.ufug.2017.10.007.

Grappling with the COVID-19 pandemic: Amanda Little, "Hunger Is Getting Worse Since the Pandemic," *Bloomberg* (June 23, 2022), https://www.bloomberg.com/opinion/articles/2022-06-23/hunger-is-worse-now-than-during-the-pandemic.

further exacerbated by the war: Pamela Falk, "150 Million More Going Hungry Worldwide as War, COVID and Climate Crisis Fuel a Global Food Emergency," *CBS News* (July 7, 2022), https://www.cbsnews.com/news/food-emergency-150-million-more-hungry-worldwide-un-food-security-report/.

Christianna Silva reported for *NPR:* Christianna Silva, "Food Insecurity in the U.S. by the Numbers," *NPR,* September 27, 2020, https://www.npr.org/2020/09/27/912486921/food-insecurity-in-the-u-s-by-the-numbers.

71   **"The real appeal of foraging is a sort of self-sufficiency":** Marie Viljoen, interview with author, September 22, 2021, New York, NY.

73   **In 2020, researchers at McGill:** Vincent Fugère, Marie-Pier Hébert, Naíla Barbosa Da Costa, Charles C.Y. Xu, Rowan D. H. Barrett, Beatrix E. Beisner, Graham Bell, et al., "Community Rescue in Experimental Phytoplankton Communities Facing Severe Herbicide Pollution," *Nature Ecology and Evolution* 4, no. 4 (2020): 578–88, https://doi.org/10.1038/s41559-020-1134-5.

And a 2022 study: Marlaina S. Freisthler, C. Rebecca Robbins, Charles Benbrook, Heather M. Young, David M. Haas, Paul Winchester, and Melissa J. Perry, "Association between Increasing Agricultural Use of 2,4-D and Population Biomarkers of Exposure: Findings from the National Health and Nutrition Examination Survey, 2001–2014," *Environmental Health* 21, vol. 1 (2022), https://doi.org/10.1186/s12940-021-00815-x.

That same year: "EPA Announces Endangered Species Act Protection Policy for New

Pesticides," US EPA, January 11, 2022, https://www.epa.gov/newsreleases/epa-announces
-endangered-species-act-protection-policy-new-pesticides.

74 **This carries a different weight:** Isabela Dias, "The Chinese Exclusion Act May Be in the
Past, But Racism Still Drives Most Immigration Policies," *Mother Jones* (August 11, 2022),
https://www.motherjones.com/politics/2022/08/the-chinese-exclusion-act-may-be-in
-the-past-but-racism-still-drives-most-immigration-policies/.

75 **"I'm coming about it from a forager's point of view":** Tama Matsuoka Wong, interview
with author, March 23, 2022, Flemington, NJ.

76 **thirty-six populations of kudzu had been identified in New Jersey:** "'Vine That Ate the
South' Comes North," New Jersey Conservation Foundation, August 21, 2019, https://www
.njconservation.org/vine-that-ate-the-south-comes-north/.

One such request: Dan Tham, "How an Indian Company Is Transforming Palm Leaves
into Tableware," *CNN* (February 28, 2022), https://edition.cnn.com/2022/02/27/business
-india/bollant-industries-india-areca-tableware-srikanth-bolla-hnk-spc-intl/index.html
#:%7E:text=At%20its%20manufacturing%20unit%2C%20staff,Prahalad%20Tiffin%20
Point%2C%20across%20India.

the state's Invasive Species Strike Team: Dana DiFilippo, "In Fight against Invasive
Plants, Strike Team Stepped up When New Jersey Wouldn't," *New Jersey Monitor* (August
2022), https://newjerseymonitor.com/2022/08/03/in-fight-against-invasive-plants-strike
-team-stepped-up-when-new-jersey-wouldnt/.

78 **Plant neurobiology:** Accademia dei Georgofili, ed., *The First Symposium on Plant Neurobiology* (Plant Behavior, 2005), https://plantbehavior.org/wp-content/uploads/2015/09/Book
-of-Abstracts-PN2005.pdf.

A 2021 *BBC Science Focus Magazine*: Efraín Rivera-Serrano, "Plants: Are They Conscious?"
BBC Science Focus Magazine (February 2021), https://www.sciencefocus.com/news/plants
-are-they-conscious/.

a 2020 paper: Vicente Raja, Paula Alexandra Silva, Roghaieh Holghoomi, and Paco Calvo,
"The Dynamics of Plant Nutation," *Scientific Reports* 10, no. 1 (2020), https://doi.org/10
.1038/s41598-020-76588-z.

The first of her landmark ecological discoveries: "Research—Suzanne Simard, Author
and Professor of Forest Ecology," Suzanne Simard, Author and Professor of Forest Ecology,
April 3, 2021, https://suzannesimard.com/research/?doing_wp_cron=1685922863.41343307
49511718750000.

when she revealed: Diane Toomey, "Exploring How and Why Trees 'Talk' to Each Other,"
Yale E360, September 1, 2016, https://e360.yale.edu/features/exploring_how_and_why
_trees_talk_to_each_other.

Such a thing is possible: Suzanne W. Simard, David S. Perry, Matthew Jones, David D.
Myrold, Daniel M. Durall, and Randy Molina, "Net Transfer of Carbon between Ectomy-
corrhizal Tree Species in the Field," *Nature* 388, no. 6642 (1997): 579–82, https://doi.org
/10.1038/41557.

by what is known as "mycorrhizal networks": Stephanie Pappas, "Do Trees Really Support Each Other through a Network of Fungi?" *Scientific American* (February 13, 2023),

https://www.scientificamerican.com/article/do-trees-support-each-other-through-a
-network-of-fungi/.

In Simard's book: "*Finding the Mother Tree: Discovering the Wisdom of the Forest* by Suzanne Simard: 9780525565994," Penguin Random House, June 21, 2022, https://www.penguin randomhouse.com/books/602589/finding-the-mother-tree-by-suzanne-simard/.

79   **A 2021 *Scientific American* interview:** Richard Schiffman, "'Mother Trees' Are Intelligent: They Learn and Remember," *Scientific American* (May 4, 2021), https://www.scientificameri can.com/article/mother-trees-are-intelligent-they-learn-and-remember/.

When she first published: Kerry Banks, "Suzanne Simard Overcame Adversity to Un-lock the Secret World of Trees," University Affairs, March 24, 2021, https://www.univer sityaffairs.ca/features/feature-article/suzanne-simard-overcame-adversity-to-unlock -the-secret-world-of-trees/.

80   **Critics of the idea of plant consciousness:** Taiz, Lincoln, Daniel L. Alkon, Andreas Draguhn, Angus S. Murphy, Michael R. Blatt, Chris Hawes, Gerhard Thiel, and David Rob-inson, "Plants Neither Possess nor Require Consciousness." *Trends in Plant Science* 24, no. 8 (2019): 677–87, https://doi.org/10.1016/j.tplants.2019.05.008.

Forests do not behave as: Ferris Jabr, "The Social Life of Forests," *New York Times* (December 7, 2020), https://www.nytimes.com/interactive/2020/12/02/magazine/tree-communication -mycorrhiza.html#:%7E:text=The%20most%20radical%20interpretation%20of,says%20 in%20her%20TED%20Talk.

7. A HOUSE IS BUT A HOME

81   **"My mom was not happy when she learned about it":** Kyle Schumann, interview with author, January 27, 2022, Charlottesville, VA.

82   **"I'd make personalized versions of board games":** Katie MacDonald, interview with au-thor, January 27, 2022, Charlottesville, VA.

In December 2017: Eric Levenson, "Thomas Fire, Once Largest in California History, Is Now 100% Contained," *CNN* (January 12, 2018), https://edition.cnn.com/2018/01/12/us/thomas -fire-california-contained.

before triggering a series: Max Golembo and Morgan Winsor, "Southern California Wild-fires Paved the Way for Deadly Mudslides," *ABC News* (January 12, 2018), https://abcnews .go.com/US/southern-california-wildfires-paved-deadly-mudslides/story?id=52257400.

83   **It's evident in Europe:** Andrew Freedman, "Heat Waves Topple Monthly, All-Time Rec-ords from Japan to Italy," *Axios* (June 30, 2022), https://www.axios.com/2022/06/30/heat -waves-europe-japan-records.

summer heat waves: Ivana Saric, Andrew Freedman, and Jacob Knutson, "Heat Wave Kills More than 2,000 People in Spain and Portugal," *Axios* (July 21, 2022), https://www.axios.com /2022/07/18/heat-wave-europe-death-toll.

It's omnipresent among: Ranj Alaaldin, "Climate Change May Devastate the Middle East. Here's How Governments Should Tackle It," *Brookings* (March 18, 2022), https://www

.brookings.edu/blog/planetpolicy/2022/03/14/climate-change-may-devastate-the-middle-east-heres-how-governments-should-tackle-it/.

the world's worst water crises: Rutger Willem Hofste, Paul Reig, and Leah Schleifer, "17 Countries, Home to One-Quarter of the World's Population, Face Extremely High Water Stress," *World Resources Institute* (August 2019), https://www.wri.org/insights/17-countries-home-one-quarter-worlds-population-face-extremely-high-water-stress.

It's unmissable across: Australia's National Science Agency, "Climate Change in Australia," CSIRO, n.dhttps://www.csiro.au/en/research/environmental-impacts/climate-change/climate-change-information.

Millions are already being forced: Abrahm Lustgarten, "Where Will Everyone Go?" *ProPublica* (July 23, 2020), https://features.propublica.org/climate-migration/model-how-climate-refugees-move-across-continents/.

People of color: Jeremy Williams, "Why Climate Change Is Inherently Racist," *BBC Future* (May 31, 2022), https://www.bbc.com/future/article/20220125-why-climate-change-is-inherently-racist.

A 2022 study: Alique Berberian, David Gonzalez, and Lara Cushing, "Racial Disparities in Climate Change-Related Health Effects in the United States," *Current Environmental Health Reports* 9, no. 3 (2022): 451–64, https://doi.org/10.1007/s40572-022-00360-w.

Whether you're in California: Winston Choi-Schagrin and Elena Shao, "Why Does the American West Have So Many Wildfires?" *New York Times* (August 3, 2022), https://www.nytimes.com/2022/08/01/climate/wildfire-risk-california-west.html.

or Florida, where current: Tim Robustelli and Shahin Vassigh, "How Miami Can Survive Climate Change," *Slate Magazine* (May 12, 2022), https://slate.com/technology/2022/05/miami-climate-change-survival.html.

Nearly all industries: Renee Cho, "How Climate Change Impacts the Economy," *State of the Planet* (June 2019), https://news.climate.columbia.edu/2019/06/20/climate-change-economy-impacts/.

84   **A 2019 study in *Nature Climate Change*:** Jeremy Martinich and Allison Crimmins, "Climate Damages and Adaptation Potential across Diverse Sectors of the United States," *Nature Climate Change* 9, no. 5 (2019): 397–404, https://doi.org/10.1038/s41558-019-0444-6.

leading fossil fuel companies: Matthew Taylor and Jonathan Watts, "Revealed: The 20 Firms behind a Third of All Carbon Emissions," *Guardian* (August 25, 2021), https://www.theguardian.com/environment/2019/oct/09revealed-20-firms-third-carbon-emissions.

industries such as building and construction: Gautam Naik, "Global Emissions from Buildings, Construction Climb to Record Levels," *Bloomberg.com* (November 9, 2022), https://www.bloomberg.com/news/articles/2022-11-09/global-emissions-from-buildings-construction-climb-to-record-levels.

the built environment makes up: "The Construction Industry Remains Horribly Climate-Unfriendly," *The Economist* (August 16, 2022), https://www.economist.com/finance-and-economics/2022/06/15/the-construction-industry-remains-horribly-climate-unfriendly.

A deciduous tree that originated: "Tree of Heaven," The Nature Conservancy, July 6, 2020,

https://www.nature.org/en-us/about-us/where-we-work/united-states/indiana/stories
-in-indiana/journey-with-nature-tree-of-heaven/.

85 **First appearing in the:** Amy Duke, "Spotted Lanternfly Lore: Penn State Experts Clear up
Falsehoods about Pest," Penn State University, January 17, 2022, https://www.psu.edu/news
/agricultural-sciences/story/spotted-lanternfly-lore-penn-state-experts-clear-falsehoods
-about-pest/.

leaving behind a sugary waste: Tawny Simisky, "Spotted Lanternfly," University of Massachusetts Amherst Center for Agriculture, Food, and the Environment, October 27, 2022, https://
ag.umass.edu/landscape/fact-sheets/spotted-lanternfly#:%7E:text=Damage,growth%20of
%20trees%20and%20shrubs.

The USDA National Invasive Species Information Center: "Spotted Lanternfly," National
Invasive Species Information Center, USDA, December 2018, https://www.invasivespe
ciesinfo.gov/terrestrial/invertebrates/spotted-lanternfly.

Hale wrote that the: Frank A. Hale, "The Invasive Spotted Lanternfly Is Spreading across
the Eastern US—Here's What You Need to Know about This Voracious Pest," *The Conversation* (July 28, 2021), https://theconversation.com/the-invasive-spotted-lanternfly-is
-spreading-across-the-eastern-us-heres-what-you-need-to-know-about-this-voracious
-pest-162919.

Last updated on December 1, 2022: New York State Integrated Pest Management, "NY-
SIPM Interactive Spotted Lanternfly Map," Looker Studio, December 1, 2022, https://data
studio.google.com/reporting/b0bae43d-c65f-4f88-bc9a-323f3189cd35/page/QUCkC.

86 **It's a little bizarre:** "History and Use of Kudzu in the Southeastern United States," Alabama Cooperative Extension System, March 8, 2022, https://www.aces.edu/blog/topics
/forestry-wildlife/the-history-and-use-of-kudzu-in-the-southeastern-united-states/.

88 **"I felt, when I came back to Hawai'i":** Joey Valenti, interview with author, October 6, 2021,
Wahiawa, HI.

89 **Hawai'i has also experienced:** Sean Connelly, "Urbanism as Island Living in Honolulu,"
Honolulu Civil Beat (August 2014), https://www.civilbeat.org/2014/08/urbanism-as-island
-living-in-honolulu/.

Since the 1800s, steel and concrete: Cheryl Smith, "Historic Architecture of Hawaii,"
Smith Brothers Construction (blog), June 9, 2021, https://smithbrothersconstruction.com
/historic-architecture-of-hawaii/.

Hawai'i is one of the most: Hugo Cox, "Dreams Adrift: Pandemic Relocations Deepen
Hawaii's Housing Crisis," *Financial Times* (July 15, 2022), https://www.ft.com/content/177
be96b-737e-469f-89be-8ba4bbfe23e0.

an albizia removal at the University of Hawai'i at Mānoa's arboretum: Clayton Truscott,
"The Albizia Project: Seeing the Wood for the Trees," *Section Magazine* (September 14,
2021), https://thesectionmag.com/2021/9/14/design/the-albizia-project.

90 **A haunted legacy in Hawai'i:** Paul Jebara and Paul Jebara, "Architecture Grad Makes Affordable Prefab Homes from Hawaii's Invasive Trees," *Dezeen* (March 18, 2020), https://
www.dezeen.com/2019/12/28/albizia-low-cost-affordable-housing-hawaii-joey-va
lenti/.

Much like kudzu, it was: Timothy A. Schuler, "Local Trees Can Help Solve Our Housing Crisis," *Hawaii Business Magazine* (May 29, 2021), https://www.hawaiibusiness.com/albizia -project/.

8. ROAD TO RECOVERY

95 **"We were looking for something":** Darryl Wilson, interview with author, September 15, 2021, Thurmond, North Carolina.

99 **In China and Japan:** "History and Use of Kudzu in the Southeastern United States," Alabama Cooperative Extension System, March 8, 2022, https://www.aces.edu/blog/topics /forestry-wildlife/the-history-and-use-of-kudzu-in-the-southeastern-united-states/.

Records in Chinese: Ka Sing Wong, George Q. Li, Kong M. Li, Valentina Razmovski-Naumovski, and Kelvin K. W. Chan, "Kudzu Root: Traditional Uses and Potential Medicinal Benefits in Diabetes and Cardiovascular Diseases," *Journal of Ethnopharmacology* 134, no. 3 (2011): 584–607, https://doi.org/10.1016/j.jep.2011.02.001.

100 **The oldest-known Chinese written reference:** Ka Sing Wong, George Q. Li, Kong M. Li, Valentina Razmovski-Naumovski, and Kelvin K. W. Chan, "Kudzu Root: Traditional Uses and Potential Medicinal Benefits in Diabetes and Cardiovascular Diseases," *Journal of Ethnopharmacology* 134, no. 3 (2011): 584–607, https://doi.org/10.1016/j.jep.2011.02.001.

It has been commonly applied in a clinical fashion: "Kudzu—Health Information Library," PeaceHealth, April 14, 2015, https://www.peacehealth.org/medical-topics/id/hn -2119009.

A 2011 study: Kai He, Xuegang Li, Xin Chen, Xiaoli Ye, Jing Huang, Ya-Nan Jin, Panpan Li, et al., "Evaluation of Antidiabetic Potential of Selected Traditional Chinese Medicines in STZ-Induced Diabetic Mice," *Journal of Ethnopharmacology* 137, no. 3 (2011): 1135–42, https://doi.org/10.1016/j.jep.2011.07.033.

In 2016, a *Laboratory Animal Research* paper: Seung-Ho Choi, Jong Inn Woo, Yeong-Su Jang, Ju-Hee Kang, Jung Eun Jang, Tae-Hoo Yi, Sang-Yong Park, Sun Yeou Kim, Yeo Sung Yoon, and Seung M. Oh, "Fermented *Pueraria Lobata* Extract Ameliorates Dextran Sulfate Sodium-Induced Colitis by Reducing pro-Inflammatory Cytokines and Recovering Intestinal Barrier Function," *Laboratory Animal Research* 32, no. 3 (2016): 151, https://doi .org/10.5625/lar.2016.32.3.151.

dextrose: Rachel Nall, "Dextrose," *Healthline* (December 21, 2021), https://www.healthline .com/health/dextrose#:~:text=Dextrose%20is%20the%20name%20of,Dextrose%20also %20has%20medical%20purposes.

a species of lactic acid bacteria: Paula Teixeira, "LACTOBACILLUS | Lactobacillus Brevis," *Elsevier eBooks* (1999): 1144–51, https://doi.org/10.1006/rwfm.1999.0900.

And a 2012 study: Seong Eun Jin, You Lee Son, Byung Soh Min, Hyun Suk Jung, and Jae Sue Choi, "Anti-Inflammatory and Antioxidant Activities of Constituents Isolated from Pueraria Lobata Roots," *Archives of Pharmacal Research* 35, no. 5 (2012): 823–37, https://doi.org /10.1007/s12272-012-0508-x.

The extract of the plant: Nada A Helal, Heba A. Eassa, Ahmed Amer, Mohamed A El-tokhy, Ivan O. Edafiogho, and Mohamed Nounou, "Nutraceuticals' Novel Formulations: The Good, the Bad, the Unknown and Patents Involved," *Recent Patents on Drug Delivery & Formulation* 13, no. 2 (2019): 105–56, https://doi.org/10.2174/1872211313666190503112040.

Not approved by governing agencies: Office of the Commissioner, "Dietary Supplements," US Food and Drug Administration, December 2021, https://www.fda.gov/consumers/consumer-updates/dietary-supplements.

In 2021, Christie Aschwanden: Christie Aschwanden, "Prohibited, Unlisted, Even Dangerous Ingredients Turn up in Dietary Supplements," *Washington Post* (June 30, 2021), https://www.washingtonpost.com/health/contaminated-supplements-unexpected-ingredients/2021/06/25/5d2227ec-bd62-11eb-83e3-0ca705a96ba4_story.html.

Dating back to the sixteenth-century days: Stephen J. Greenblatt and John S. Morrill, "Elizabeth I," *Encyclopaedia Britannica* (June 2, 2023).

101   **"I was doing some reading and came across this article":** Scott Lukas, interview with author, September 17, 2021, Belmont, MA.

The doctor had stumbled upon: Wing-Ming Keung and Bert L. Vallee, "Daidzin and Daidzein Suppress Free-Choice Ethanol Intake by Syrian Golden Hamsters," *Proceedings of the National Academy of Sciences of the United States of America* 90, no. 21 (1993): 10008–12, https://doi.org/10.1073/pnas.90.21.10008.

Google Scholar wasn't established: Devon Delfino, "What Is Google Scholar? How to Use the Academic Database for Research," *Business Insider* (August 2022), https://www.businessinsider.com/guides/tech/google-scholar?international=true&r=US&IR=T.

105   **One example is a 2015 study:** David M. Penetar, Lindsay H. Toto, David J. Lee, and Scott E. Lukas, "A Single Dose of Kudzu Extract Reduces Alcohol Consumption in a Binge Drinking Paradigm," *Drug and Alcohol Dependence* 153 (August 2015): 194–200, https://doi.org/10.1016/j.drugalcdep.2015.05.025.

106   **An earlier study, first published online in 2012:** Scott E. Lukas, David M. Penetar, Zhaohui Su, Thomas Geaghan, Melissa A. Maywalt, Michael C. Tracy, John Rodolico, Christopher Palmer, Zhongze Ma, and David J. Lee, "A Standardized Kudzu Extract (NPI-031) Reduces Alcohol Consumption in Nontreatment-Seeking Male Heavy Drinkers," *Psychopharmacology* 226 (2013): 65–73, https://doi.org/10.1007/s00213-012-2884-9.

107   **Roughly 98% of publications in science:** Valeria Ramírez-Castañeda, "Disadvantages in Preparing and Publishing Scientific Papers Caused by the Dominance of the English Language in Science: The Case of Colombian Researchers in Biological Sciences," *PLOS One* 15, no. 9 (September 16, 2020): e0238372, https://doi.org/10.1371/journal.pone.0238372.

9. GOT GOATS?

109   **"Here in the US":** Qinfeng Guo, interview with author, July 6, 2022, Research Triangle Park, NC.

In 2011, Guo coauthored a paper: Zhenyu Li, Quanfeng Dong, Thomas P. Albright, and Qinfeng Guo, "Natural and Human Dimensions of a Quasi-Wild Species: The Case of Kudzu," *Biological Invasions* 13, no. 10 (2011): 2167–79, https://doi.org/10.1007/s10530-011-0042-7.

110   **A 2010 paper published on** *J-STAGE*: Misako Ito and Institute for Urban Weed Science, "Weed Introduction Series: Kudzu (Pueraria Lobata Ohwi)," *J-STAGE* 2, no. 36 (2010): 41, https://www.jstage.jst.go.jp/article/iuws/2/0/2_36/_pdf.

       the potential of kudzu: "NASA Technical Reports Server (NTRS) 19800009283: Energy from Aquatic Plant Wastewater Treatment Systems: NASA Technical Reports Server (NTRS): Free Download, Borrow, and Streaming," Internet Archive, September 1, 1979, https://archive.org/details/NASA_NTRS_Archive_19800009283/page/n1/mode/2up.

       Critics of the push: Michael Todd, "Not Ready for Prime Time: Making Fuel Out of Invasive Plants," *Pacific Standard* (June 14, 2017), https://psmag.com/environment/ready-prime-time-making-fuel-invasive-plants-70402.

111   **Although more than 1.9 million:** Dipika Kadaba and John R. Platt, "A Steal of a Deal: How Ranchers Take Advantage of Public Lands," *The Revelator* (July 2020), https://therevelator.org/cattle-public-lands/.

       but hold no "rights": Erik Molvar, "Livestock Grazing on Federal Public Lands Is a Privilege —Not a Right," *The Hill* (April 24, 2018), https://thehill.com/opinion/energy-environment/384270-livestock-grazing-on-federal-public-lands-is-a-privilege-not-a/.

       According to a 1991 report: E.G. Rhoden, "Kudzu as a Feed for Angora Goats," US Forest Service Research and Development, 1991, https://www.fs.usda.gov/treesearch/pubs/6751.

       And a 2019 article: "Kudzu (Pueraria Lobata)," Mississippi State University Extension Service, n.d., http://www.ext.msstate.edu/content/kudzu-pueraria-lobata.

       It referenced a 1947 federal experiment: Emery A. Telford, "Tropical Kudzu in Puerto Rico," US Department of Agriculture, 1947, Internet Archive, https://archive.org/details/CAT31289488/page/10/mode/2up?ref=ol&view=theater.

112   **Outlets like** *BBC News:* Joanna Jolly, "The Goats Fighting America's Plant Invasion," *BBC News* (January 13, 2015), https://www.bbc.com/news/magazine-30583512.

       the *Washington Post:* Fritz Hahn, "The Goat Mowers Are Coming Back to Congressional Cemetery," *Washington Post* (July 14, 2015), https://www.washingtonpost.com/news/going-out-guide/wp/2015/07/14/the-goat-mowers-are-coming-back-to-congressional-cemetery/.

       and *National Geographic:* Melody Kramer, "The Kids Are Alright: Goats That Double as Lawnmowers," *National Geographic* (May 3, 2021), https://www.nationalgeographic.com/science/article/130808-goats-lawnmowers-environment-cemetery-congressional-washington-dc.

       One of those is: "Goats v. Kudzu," *Post and Courier* (February 10, 2020), https://www.postandcourier.com/berkeley-independent/news/goats-v-kudzu/article_058438f2-da9a-573c-9e2f-8e1f50a94f93.html.

       First launched in 2014: "Goat Powered Invasive Removal," Clemson University College of Agriculture, Forestry and Life Sciences, April 22, 2022, https://www.clemson.edu/cafls/research/hunnicutt/invasive/goats.html.

       "armed with loppers and hedge trimmers": Phil Pierce, "Alabama Communities Partner to Restore Their Watersheds," *Alabama News Center* (July 18, 2018), https://alabamanewscenter.com/2018/07/11/alabama-communities-partner-to-restore-their-watersheds/.

       They are merely one of: "A Little Sheep Goes a Long Way in Managing Kudzu," News Center, Georgia Institute of Technology, November 6, 2014, https://news.gatech.edu/news/2014/11/06/little-sheep-goes-long-way-managing-kudzu.

113   **Livestock can cause lasting:** Mohamed M. Abdalla, Annette Hastings, David R. Chadwick, David R. Jones, Chris Evans, Michael P. Jones, Robert C. Rees, and Pete Smith, "Critical Review of the Impacts of Grazing Intensity on Soil Organic Carbon Storage and Other Soil Quality Indicators in Extensively Managed Grasslands," *Agriculture, Ecosystems & Environment* 253 (February 2018): 62–81, https://doi.org/10.1016/j.agee.2017.10.023.

This is only one of: Jennifer M. Schieltz and Daniel I. Rubenstein. "Evidence Based Review: Positive versus Negative Effects of Livestock Grazing on Wildlife. What Do We Really Know?" *Environmental Research Letters* 11, no. 11, (2016): 113003, https://doi.org/10.1088/1748-9326/11/11/113003.

in a body of literature: Nicolas Galleguillos, Keefe Keeley, and Stephen J. Ventura, "Assessment of Woodland Grazing in Southwest Wisconsin," *Agriculture, Ecosystems & Environment* 260 (June 2018): 1–10, https://doi.org/10.1016/j.agee.2018.03.012.

according to a 2016 *Grist* news article: Ben Adler, "Cattle Grazing Is a Climate Disaster, and You're Paying for It," *Grist* (April 1, 2021), https://grist.org/climate-energy/cattle-grazing-is-a-climate-disaster-and-youre-paying-for-it/.

Despite the downsides: Ronald A. Rathfon, Skye M. Greenler, and Michael W. Jenkins, "Effects of Prescribed Grazing by Goats on Non-native Invasive Shrubs and Native Plant Species in a Mixed-hardwood Forest," *Restoration Ecology* 29, no. 4 (2021), https://doi.org/10.1111/rec.13361.

according to a Purdue University: Wendy Mayer, "Goat Grazing Could Be Option for Invasive Species Removal," Purdue University College of Agriculture, September 29, 2021, https://ag.purdue.edu/news/department/forestry-and-natural-resources/2021/09/goat-grazing-could-be-option-for-invasive-species-removal.html.

In Burlington, Wisconsin: Michele Arduengo, "Just in Time for Wisconsin's Invasive Species Month: Goats," Promega Connections, June 21, 2017, https://www.promegaconnections.com/just-in-time-for-wisconsins-invasive-species-month-goats/.

actively being used for land management: The Green Goats, "The Green Goats," Facebook, n.d., https://www.facebook.com/TheGreenGoats/?hc_ref=PAGES_TIMELINE&fref=nf.

Chloe Lewis at the *Highland Echo*: Chloe Lewis, "Goats Kill Kudzu in MC Woods," *Highland Echo* (October 19, 2021), https://highlandecho.com/goats-kill-kudzu-in-mc-woods/.

Over in North Carolina, Madison Elliot: Madison Elliott, "City of Belmont Hires Goats to Help Eat Invasive Species of Plant," *Spectrum News 1 Charlotte* (June 6, 2022), https://spectrumlocalnews.com/nc/charlotte/news/2022/06/03/city-of-belmont-hires-goats-to-help-eat-invasive-species-of-plant.

114   **"The use of invasive plants for alternative purposes like this":** Stephen Enloe, interview with author, October 4, 2021, Gainesville, Florida.

In the 1900s, the USDA promoted the tallow tree: "Southeast Exotic Pest Plant Council Invasive Plant Manual: Chinese Tallowtree," n.d., Southeast Exotic Pest Plant Council, https://www.se-eppc.org/manual/sase.html#:%7E:text=Chinese%20tallowtree%20is%20native%20to,establish%20a%20soap%20making%20industry.

Today, the tallow tree: Amy Vu, "What Is Happening with the Proposal for the Chinese Tallow Tree (Triadica Sebifera) Insect Release?" University of Florida IFAS Entomology and Nematology Department, February 16, 2021, https://blogs.ifas.ufl.edu/entnemdept/2021

/02/16/what-is-happening-with-the-proposal-for-the-chinese-tallow-tree-triadica-sebi
fera-insect-release/.

10. MISSING IN ACTION

116 **With that, my desire:** Beth Lebwohl, "How Soil Shatters like Glass, and What That Means,"
EarthSky (December 29, 2010), https://earthsky.org/earth/how-soil-shatters-like-glass-and
-what-that-means-for-climate-change/.

118 **A 2001 chapter penned:** Richard J. Blaustein, "Kudzu's Invasion into Southern United
States Life and Culture," US Forest Service Research and Development, 2001, https://www
.fs.usda.gov/research/treesearch/6184.

Chris Park and Michael Allaby's: Chris Park and Michael Allaby, *A Dictionary of Environment
and Conservation,* 3rd ed. (Oxford University Press eBooks, 2017), https://doi.org/10.1093
/acref/9780191826320.001.0001.

"This is hard to determine as it depends on the scales to measure the range": Qinfeng
Guo, interview with author, July 6, 2022, Research Triangle Park, NC.

119 **"Our program only measures kudzu on forestland":** Sonja Oswalt and Christopher Oswalt, email interview with author, September 23, 2021.

The two contributed to: Sonja N. Oswalt, Chris Oswalt, Alycia Crall, Robert J. Rabaglia,
Michael W. Schwartz, and Becky K. Kerns, "Inventory and Monitoring of Invasive Species,"
Springer eBooks (2021): 231–42, https://doi.org/10.1007/978-3-030-45367-1_10.

120 **A 2015 Congressional Research Service report:** "Invasive Species: Control Options and Issues for Congress," *CRS Reports* (Congressional Research Service, 2015), https://www.every
crsreport.com/reports/R44011.html.

121 **Founded in 1999:** "SE-EPPC—Southeast Exotic Pest Plant Council," University of Georgia
Center for Invasive Species and Ecosystem Health, n.d., https://www.se-eppc.org/.

122 **"A number of things happened":** Stephen Enloe, interview with author, October 4, 2021,
Gainesville, FL.

123 **My quest for answers:** "EDDMapS," University of Georgia Center for Invasive Species and
Ecosystem Health, n.d., https://www.eddmaps.org/.

The numbers are difficult: "Sustainability Report for Georgia's Forests: January 2019," Georgia Forestry Commission, 2019, https://gatrees.org/wp-content/uploads/2020/01/Sustain
ability-Report-for-Georgias-Forests-January-2019-WEB.pdf#:~:text=Georgia's%20
forest%20area%20has%20remained,other%20state%20in%20the%20nation.

124 **After reaching out to the agency:** "Georgia Forestry Commission," State of Georgia, n.d.,
https://gatrees.org/about/.

"There really isn't very much research on seed fertility and fecundity": Nancy Loewenstein, interview with author, August 27, 2021, Auburn, AL.

125 **a paragraph that quotes directly from a 2015 *Smithsonian* article:** Bill Finch, "The True
Story of Kudzu, the Vine That Never Truly Ate the South," *Smithsonian Magazine* (August
24, 2015), https://www.smithsonianmag.com/science-nature/true-story-kudzu-vine-ate
-south-180956325/.

127 **She shares a resource published in 2013:** Johnnie L. Gentry, George P. Johnson, Brent T. Baker, C. Theo Witse, and Jennifer D. Og, "Atlas of the Vascular Plants of Arkansas" (978-0-615-67980-8), The Arkansas Vascular Flora Committee, University of Arkansas, 2013, https://www.uark.edu/~arkflora.

a geological formation made up of narrow hills: "Crowley's Ridge," Encyclopedia of Arkansas, November 22, 2022, https://encyclopediaofarkansas.net/entries/crowleys-ridge-12/.

128 **Listed by the Florida Exotic Pest Plant Council:** Tom Palmer, "Now in Polk, Kudzu Weed 'Category 1' Invasive Species," *The Ledger* (August 26, 2008), https://www.theledger.com/story/news/2008/08/26/kudzu-infestation-spreads-to-polk/25946163007/.

Florida's Invasive Plant Management: "Invasive Plant Management," Florida Fish and Wildlife Conservation Commission, n.d., https://myfwc.com/wildlifehabitats/habitat/invasive-plants/.

But according to the University of South Florida's Institute for Systematic Botany: Institute for Systematic Botany. "Pueraria Montana Var. Lobata—Species Details," Atlas of Florida Plants, n.d., https://florida.plantatlas.usf.edu/Plant.aspx?id=432.

129 **She is the lead author:** Paulina Harron, Omkar Joshi, Christopher B. Edgar, Shishir Paudel, and Arjun S. Adhikari, "Predicting Kudzu (Pueraria Montana) Spread and Its Economic Impacts in Timber Industry: A Case Study from Oklahoma," *PLOS ONE* 15, no. 3 (2020): e0229835, https://doi.org/10.1371/journal.pone.0229835.

The most up-to-date cost: Kerry O. Britton, David Orr, Jianghua Sun, and USDA Forest Service Publication, "Chapter 25 Kudzu—Biological Control of Invasive Plants in the Eastern United States," The Bugwood Network, 2002, https://www.invasive.org/biocontrol/25Kudzu.cfm.

A 1993 study by the Congressional Office of Technology Assessment: Roy Van Driesche, "Biological Control of Invasive Plants in the Eastern United States," AGRIS: International Information System for the Agricultural Science and Technology, 2002, https://agris.fao.org/agris-search/search.do?recordID=US201300081351.

In 2020, the US Department of the Interior earmarked $143 million: Jean E. Fantle-Lepczyk, Phillip J. Haubrock, Andrew A. Kramer, Tatenda Dalu, Anna J. Turbelin, Robert Crystal-Ornelas, Christophe Diagne, and Franck Courchamp, "Economic Costs of Biological Invasions in the United States," *Science of the Total Environment* 806 (February 2022): 151318, https://doi.org/10.1016/j.scitotenv.2021.151318.

Alone, kudzu packs a pricey punch: Maryam Shahrtash and Shawn T. Brown, "Drivers of Foliar Fungal Endophytic Communities of Kudzu (Pueraria Montana Var. Lobata) in the Southeast United States," *Diversity* 12, no. 5 (2020): 185, https://doi.org/10.3390/d12050185.

"These economic impacts definitely serve as an incentive for governments": Paulina Harron, interview with author, June 25, 2021, Tarrytown, NY.

130 **A 2021 study in *Remote Sensing*:** "Based Invasive Kudzu Mapping: A Case Study in Knox County, Tennessee," *Remote Sensing* 13, no. 22 (November 12, 2021): 4551. https://doi.org/10.3390/rs13224551.

131 **And a 2020 paper:** Liang, Wanwan, Mongi A. Abidi, Luis Carrasco, J. J. McNelis, Liem Tran, Yingkui Li, and Jerome F. Grant. "Mapping Vegetation at Species Level with High-Resolution

Multispectral and Lidar Data Over a Large Spatial Area: A Case Study with Kudzu." *Remote Sensing* 12, no. 4 (February 12, 2020): 609. https://doi.org/10.3390/rs12040609.

As of 2022, eleven states in the US have a law against: "Early Detection & Distribution Mapping System (EDDMapS)," University of Georgia Center for Invasive Species and Ecosystem Health, n.d., accessed August 15, 2022, https://www.eddmaps.org/.

132   **"The biggest obstacle is that when you do what we do":** Chuck Bargeron, interview with author, October 19, 2021, Tifton, GA.

134   **In 2016, the legislation was updated:** "Executive Order—Safeguarding the Nation from the Impacts of Invasive Species," The White House (press release), December 5, 2016, https://obamawhitehouse.archives.gov/the-press-office/2016/12/05/executive-order-safeguarding-nation-impacts-invasive-species.

Published in spring 2022: "National Invasive Species Council Annual Work Plan FY 2022," National Invasive Species Council, US Department of the Interior, n.d., https://www.doi.gov/sites/doi.gov/files/nisc-fy2022-annual-wp-march312022.pdf.

From 1999 to 2019: "National Invasive Species Council," US Department of the Interior, April 5, 2023, https://www.doi.gov/invasivespecies/about-isac.

The funding was cut: "Interior Department Calls for Nominations to Serve on Committee Coordinating Federal Actions on Invasive Species," US Department of the Interior, January 24, 2022, https://www.doi.gov/pressreleases/interior-department-calls-nominations-serve-committee-coordinating-federal-actions.

the US Department of the Interior announced: "Interior Department Calls for Nominations to Serve on Committee Coordinating Federal Actions on Invasive Species," US Department of the Interior, January 24, 2022, https://www.doi.gov/pressreleases/interior-department-calls-nominations-serve-committee-coordinating-federal-actions.

Two weeks later: "USDA Provides More Than $70 Million to Protect Crops and Natural Resources from Invasive Pests and Diseases in 2022," USDA APHIS, February 8, 2022, https://www.aphis.usda.gov/aphis/newsroom/news/sa_by_date/sa-2022/ppa-7721#:%7E:text=WASHINGTON%2C%20February%2008%2C%202022%20%E2%80%94,and%20threat%20mitigation%3B%20to%20safeguard.

135   **In late 2022:** "Advisory Committee Charts a Path Forward for Controlling Destructive Invasive Species," US Department of the Interior (press release), March 8, 2023, https://www.doi.gov/pressreleases/advisory-committee-charts-path-forward-controlling-destructive-invasive-species.

Bargeron is among the: "National Invasive Species Council," US Department of the Interior, April 5, 2023, https://www.doi.gov/invasivespecies/about-isac.

The emerald ash borer: Adrian Higgins, "How an Uninvited Pest Doomed the Ash Tree," *Washington Post* (June 19, 2018), https://www.washingtonpost.com/lifestyle/home/an-invasive-beetle-has-created-a-nightmare-on-ash-street/2018/06/18/247dbcd2-6f48-11e8-bf86-a2351b5ece99_story.html.

before spreading to at least: "Emerald Ash Borer," USDA APHIS, March 27, 2023, https://www.aphis.usda.gov/aphis/ourfocus/planthealth/plant-pest-and-disease-programs/pests-and-diseases/emerald-ash-borer.

killing tens of millions: Kristine Grayson, "The Invasive Emerald Ash Borer Has Destroyed Millions of Trees—Scientists Aim to Control It with Tiny Parasitic Wasps," *The Conversation* (August 27, 2021), https://theconversation.com/the-invasive-emerald-ash-borer -has-destroyed-millions-of-trees-scientists-aim-to-control-it-with-tiny-parasitic-wasps -158403.

136   **Established in the summer of 2022:** "Southeast RISCC," Center for Invasive Species and Ecosystem Health at the University of Georgia College of Agricultural and Environmental Sciences and Warnell School of Forestry and Natural Resources, June 9, 2023, https:// southeastriscc.org/#:~:text=Southeast%20RISCC%20Goals%20%26%20Research,via%20 information%20sharing%20and%20research.

The National Association of Invasive Plant Councils: "NAEPPC—National Association of IPCs," Center for Invasive Species and Ecosystem Health at the University of Georgia, n.d., https://www.na-ipc.org/.

The North American Invasive Species Management Association: "NAISMA," North American Invasive Species Management Association, 2023, https://naisma.org/.

II. CARBON CONUNDRUM

138   **"As a plant person, you see trees and you see bushes":** Lewis Ziska, interview with author, May 13, 2022, New York, NY; Lewis H. Ziska, *Greenhouse Planet: How Rising Co2 Changes Plants and Life as We Know It,* Columbia University Press, 2022.

142   **Plants can even contain:** Jerald Pinson, "Ferns Finally Get a Genome, Revealing a History of DNA Hoarding and Kleptomania," Florida Museum of Natural History, September 1, 2022, https://www.floridamuseum.ufl.edu/science/ferns-finally-get-a-genome-revealing-a -history-of-dna-hoarding-and-kleptomania/.

143   **"We just don't know":** Qinfeng Guo, interview with author, July 6, 2022, Research Triangle Park, NC.

One can just look at: Sandra L. Hoffberg and Rodney Mauricio, "The Persistence of Invasive Populations of Kudzu near the Northern Periphery of Its Range in New York City Determined from Historical Data," *Journal of the Torrey Botanical Society* 143, no. 4 (2016): 437, https://doi.org/10.3159/torrey-d-16-00032.1.

144   **In a 2014 study:** Mioko Tamura and Nishanth Tharayil, "Plant Litter Chemistry and Microbial Priming Regulate the Accrual, Composition and Stability of Soil Carbon in Invaded Ecosystems," *New Phytologist* 203, no. 1 (2014): 110–24, https://doi.org/10.1111/nph.12795.

Grist's **Jim Meyer:** Jim Meyer, "Tired of Just Ruining Your Garden, Kudzu Destroys the Climate, Too," *Grist* (July 8, 2014), https://grist.org/climate-energy/tired-of-just-ruining-your -garden-kudzu-destroys-the-climate-too/.

145   **Four years before that paper was published:** Jonathan E. Hickman, Shiliang Wu, Loretta J. Mickley, and Manuel T. Lerdau, "Kudzu (*Pueraria Montana*) Invasion Doubles Emissions of Nitric Oxide and Increases Ozone Pollution," *Proceedings of the National Academy of Sciences of the United States of America* 107, no. 22 (2010): 10115–19. https://doi.org/10.1073/pnas .0912279107.

An adjunct associate research scientist: "Kudzu Harms Air, Not Just Ecosystems, Says Study," The Earth Institute at Columbia University, May 18, 2010, https://www.earth.columbia.edu/articles/view/2695.

"It had a mythic quality to it": Jonathan Hickman, interview with author, August 19, 2021, New York, NY.

146   **The lead author of a 2016:** Sandra L. Hoffberg and Rodney Mauricio, "The Persistence of Invasive Populations of Kudzu near the Northern Periphery of Its Range in New York City Determined from Historical Data," *Journal of the Torrey Botanical Society* 143, no. 4 (2016): 437, https://doi.org/10.3159/torrey-d-16-00032.1.

147   **"I was collecting leaf tissue from kudzu and wisteria":** Sandra Hoffberg, interview with author, August 20, 2021, New York, NY.

Using data collected for a 1989 study: Edward B. Frankel, "Distribution of Pueraria Lobata in and Around New York City," *Bulletin of the Torrey Botanical Club* 116, no. 4 (1989): 390, https://doi.org/10.2307/2996629.

"When it comes to climate change": Mark Lynas, Benjamin Z. Houlton, and Simon Perry, "Greater than 99% Consensus on Human Caused Climate Change in the Peer-Reviewed Scientific Literature," *Environmental Research Letters* 16, no. 11 (2021): 114005, https://doi.org/10.1088/1748-9326/ac2966.

According to NASA: "Do Scientists Agree on Climate Change?" NASA: Global Climate Change, June 9, 2023, https://climate.nasa.gov/faq/17/do-scientists-agree-on-climate-change/.

One of the latest Intergovernmental Panel on Climate Change reports: "Climate Change 2022: Impacts, Adaptation and Vulnerability," Intergovernmental Panel on Climate Change, 2022, https://www.ipcc.ch/report/ar6/wg2/.

148   **deemed "the most authoritative reports on the topic":** Ayesha Tandon, "Analysis: How the Diversity of IPCC Authors Has Changed over Three Decades," *Carbon Brief* (March 15, 2023), https://www.carbonbrief.org/analysis-how-the-diversity-of-ipcc-authors-has-changed-over-three-decades/.

Climate modeling produced: Rosamund Pearce, "Timeline: The History of Climate Modelling," *Carbon Brief* (February 2023), https://www.carbonbrief.org/timeline-history-climate-modelling/.

149   **one prolific news article:** Bill Finch, "The True Story of Kudzu, the Vine That Never Truly Ate the South," *Smithsonian Magazine* (August 24, 2015), https://www.smithsonianmag.com/science-nature/true-story-kudzu-vine-ate-south-180956325/.

150   **"That was one of those nicknames that just became my name":** James Miller, interview with author, November 4, 2021, Auburn, AL.

151   **What is known about:** Qinfeng Guo and Becky K. Kerns, "Climate Change and Invasive Plants in Forests and Rangelands," US Department of Agriculture, Forest Service, Climate Change Resource Center, September 2012, https://www.fs.usda.gov/ccrc/topics/invasive-plants.

In a 2016 *Atlanta Journal-Constitution* article: Dan Chapman, "Kudzu's Entanglement of South Begins to Unravel," *Atlanta Journal-Constitution* (August 12, 2016), https://www.ajc.com/news/state-regional/kudzu-entanglement-south-begins-unravel/2DcsTsbnkL803xwZtvBqpO/.

152   **According to ResearchGate:** "James H. Miller Research Profile," ResearchGate, n.d., https://www.researchgate.net/profile/James-Miller-16.

        And much of that delved: Nina Keenam, "Southerners Were Paid to Plant Kudzu," *Andalusia Star-News* (June 24, 2017), https://www.andalusiastarnews.com/2017/06/23/southerners-were-paid-to-plant-kudzu/.

153   **The former is a flowering shrub:** "Callery Pear Publication Talks Treatment, Control," US Forest Service, May 25, 2021, https://www.fs.usda.gov/inside-fs/delivering-mission/deliver/callery-pear-publication-talks-treatment-control.

        invasive deciduous tree: Chelsey Cox, "These Invasive Trees Smell like Rotting Fish and Kill Plants. State Bans Want Bradford Pears Gone," *USA TODAY* (April 19, 2022), https://eu.usatoday.com/story/news/nation/2022/04/18/bradford-pear-trees-ban/7122246001/.

12. CONDEMNED

155   **"It's all mountains, the whole thing":** Casey Lance Brown, interview with author, February 10, 2022, Asheville, NC.

157   **In 1968, *The Kudzu*:** Mississippi Student News Project, "The Kudzu," Catalog—Bowling Green State University Libraries, n.d., https://maurice.bgsu.edu/record=b2652092%7ES9.

        where it ran for four years: Bill Davis, "Kudzu, Mississippi's Radical Underground Newspaper, 1968–1972," 1960s: Days of Rage, March 29, 2018, https://1960sdaysofrage.wordpress.com/2018/03/29/kudzu-mississippis-radical-underground-newspaper-1968-1972/.

        Founded by former and current: Ted Ownby, "Kudzu," *Mississippi Encyclopedia* (April 14, 2018), https://mississippiencyclopedia.org/entries/kudzu-newspaper/.

158   **Covering cases of discrimination:** Kayla Robison, "The Kudzu," Counterculture in Mississippi, December 5, 2015, https://kjrobison.wixsite.com/counterculture-ms/the-kudzu.

        In May of 1963: James Dickey, "Kudzu," *New Yorker* (May 11, 1963), https://www.newyorker.com/magazine/1963/05/18/kudzu.

        while at *Harper's Magazine*: Ben Windham, "SOUTHERN LIGHTS: In the South, Kudzu Has Got Us Covered," *Tuscaloosa News* (September 20, 2015), https://eu.tuscaloosanews.com/story/opinion/columns/2015/09/20/southern-lights-in-the-south-kudzu-has-got-us-covered/29964312007/.

        Willie Morris labeled it as "sinister": Linton Weeks, "A Southerner at Heart," *Tampa Bay Times* (December 22, 2019), https://www.tampabay.com/archive/1999/08/08/a-southerner-at-heart/.

        Marjie Short's *Kudzu*: Academy of Motion Picture Arts and Sciences. "THE 49TH ACADEMY AWARDS | 1977," n.d. https://www.oscars.org/oscars/ceremonies/1977.

        Sheets of the vine blanketed: Cody Hamman, "Trailer: Kudzu Zombies Attack in the American Southeast," JoBlo, July 31, 2021, https://www.joblo.com/trailer-kudzu-zombies-attack-in-the-american-southeast-301/.

        And in 1996: "The Amazing Story of Kudzu," Max Shores, January 29, 2020, https://maxshores.com/the-amazing-story-of-kudzu/.

That same year, Ron Harrist: Ron Harrist, "Kudzu: Dixie's Creeping Death," Associated Press, August 12, 1996, https://news.google.com/newspapers?nid=1842&dat=19960812&id=AtEyAAAAIBAJ&sjid=_8gEAAAAIBAJ&pg=1240,1729535.

159   **Hanging in the Mississippi Museum of Art:** "Kudzu," 2022. https://mma.emuseum.com/objects/583/kudzu.

In Huntsville, Alabama: "Kudzu Productions, Inc.," Huntsville/Madison County Chamber, October 7, 2019, https://cm.hsvchamber.org/list/member/kudzu-productions-inc-1010.

The vine has also made recent cinematic appearances: "UNDER THE KUDZU," AIA Film Challenge, July 21, 2018, https://aiafilmchallenge.org/video-contest/under-the-kudzu/.

"The Carrot in the Kudzu": IMDb, "'Bones' The Carrot in the Kudzu," March 24, 2014. https://www.imdb.com/title/tt3547454.

You can even catch: IMDb, "'Sweet Magnolias' Save My Place," July 20, 2023. https://www.imdb.com/title/tt22460904/.

160   **Carolina Kudzu is:** Brett Friedlander, "Legends Never Die: Sandlot Baseball Making Comeback, *North State Journal* (April 27, 2022), https://nsjonline.com/article/2022/04/legends-never-die-sandlot-baseball-making-comeback/.

the Kudzu Classic Championships: "Kudzu Classic Championships," A5 Volleyball Club, n.d., https://a5volleyball.com/view/special/view_onetourn/&tournid=9.

In Atlanta, Georgia: Derek H. Alderman, "When an Exotic Becomes Native: Taming, Naming, and Kudzu as Regional Symbolic Capital," *Southeastern Geographer* 55, no. 1 (2015): 32–56, https://doi.org/10.1353/sgo.2015.0004.

open since 1999: "Serving Children and Teens Living with Type 1 Diabetes," Camp Kudzu, October 11, 2022, https://www.campkudzu.org/about-us/#:%7E:text=Camp%20Kudzu%20was%20founded%20in,a%20fun%2C%20medically%20supervised%20environment.

In 2011, multiple news organizations: "Vine in North Carolina Town Said to Look Like Jesus," *Lubbock Avalanche-Journal* (June 29, 2011), https://www.lubbockonline.com/story/news/nation-world/2011/06/30/vine-north-carolina-town-said-resemble-jesus/15226704007/.

Chris Totten, a video game designer: Chris Totten, "Zen and the Art of Retro Level Design in 'Kudzu,'" *Medium* (December 24, 2021), https://totter87.medium.com/zen-and-the-art-of-retro-level-design-in-kudzu-81cc7545fb3e.

First premiering in the summer: Kudzu Killers Podcast, "Kudzu Killers: Homicide and Sweet Tea," Spreaker, 2020, https://www.spreaker.com/show/kudzu-killers-homicide-and-sweet-tea.

161   **And Chris Lindland:** Holly Richmond, "Hilarious Kickstarter Alert: Kudzilla, the Godzilla Made out of Invasive Kudzu Vines," *Grist* (November 4, 2013), https://grist.org/living/hilarious-kickstarter-alert-kudzilla-the-godzilla-made-out-of-invasive-kudzu-vines/.

Lindland was ultimately unsuccessful: Chris Lindland, "KUDZILLA—A Monumental Monster Made of Kudzu," Kickstarter, 2013, https://www.kickstarter.com/projects/2048863414/kudzilla-a-monumental-monster-made-of-kudzu.

13. MAN OF THE HOUR

162 **"If you hear music in the background, it's not a disco":** Derek Alderman, interview with author, May 24, 2022, Knoxville, TN.

163 **By then, he had already read the work:** John J. Winberry and David R. Jones, "Rise and Decline of the 'Miracle Vine': Kudzu in the Southern Landscape," *Southeastern Geographer* 13, vol. 2 (1973): 61–70, https://doi.org/10.1353/sgo.1973.0004.

164 **It was 2004 when:** D. H. Alderman, "Channing Cope and the Making of a Miracle Vine," *Geographical Review* 94, no. 2 (2004): 157–177, http://www.jstor.org/stable/30033969.

 Business Week even designated it: Kristen Hinman, "Kudzu—Japan's Wonder Vine," *History Net,* September 29, 2017, https://www.historynet.com/kudzu-japans-wonder-vine/.

 A 1949 *Time* article: "The Press: The Kudzu Kid," *TIME.com* (July 4, 1949), https://content .time.com/time/subscriber/article/0,33009,888578-1,00.html.

165 **That same article mentions the then-recent success of Cope's debut book:** Richard Solomon, "Kudzu Is So Much More Than the 'Vine That Ate the South,'" *Slate Magazine* (August 28, 2021), https://slate.com/news-and-politics/2021/08/kudzu-south-japan-meta phor.html.

 Alderman himself once described Cope: Derek H. Alderman, "Channing Cope and the Making of a Miracle Vine," *Geographical Review* 94, no. 2 (2004): 157–77, http://www.jstor.org /stable/30033969.

 Their sons were Channing and Willard: *Street Railway Journal* 60 (McGraw Publishing Company, 1922), https://books.google.com/books?id=5zl5AQAAMAAJ&pg=PA996&lpg=PA 996&dq=willard+cope+covington+georgia&source=bl&ots=jCkSHqa7Nv&sig=ACfU3U 2W1yi2bngOoioJ6eXF7zHknC8WKA&hl=en&sa=X&ved=2ahUKEwjl7s2a-J_zAhWS SjABHQY3DYAQ6AF6BAgNEAM#v=onepage&q=willard%20cope%20covington%20 georgia&f=false.

 She also greatly contributed: Cal Gough, "Atlanta's 'Authors Grove,'" Atlanta Booklover's Blog, October 21, 2011, https://atlantareader.wordpress.com/2011/10/21/atlantas-authors -grove/.

 The Special Collections Library: Julia Evans Cope, "Julia Evans Cope Letter," Special Collections Libraries at the University of Georgia, 1929, https://sclfind.libs.uga.edu/sclfind /view?docId=ead/ms3857.xml%3Bquery=%3Bbrand=default.

 Historical records suggest: Derek H. Alderman, "Channing Cope and the Making of a Miracle Vine," *Geographical Review* 94, no. 2 (2004): 157–77, http://www.jstor.org/stable /30033969.

 In 1927, he bought his own farm: Channing Cope, *Front Porch Farmer* (Turner E. Smith & Co., 1946), https://babel.hathitrust.org/cgi/pt?id=coo.31924003340340&view=1up&seq=9& q1=ruthie.

 Helen Athleen Rohrer: "Athleen Cope," Find a Grave Index, 1600s–Current, Ancestry.Com, n.d., https://www.ancestry.com/discoveryui-content/view/142494308:60525?tid=&pid=& queryId=08761128e3117a39179ce937b0d1cc35&_phsrc=uxA3&_phstart=successSource.

Ruth Gentry Cope: "Ruth Cope Gentry (1923–1983)," Find a Grave, n.d., https://www.find agrave.com/memorial/52020475/ruth-gentry.

166   **"A farmer is lost without an understanding wife":** Channing Cope, *Front Porch Farmer* (Turner E. Smith & Co., 1946), https://babel.hathitrust.org/cgi/pt?id=c00.31924003340340 &view=1up&seq=9&q1=ruthie.

In his prime: Clinton Crockett Peters, "The Miracle Vine," The Awl, August 29, 2017, https://www.theawl.com/2017/08/kudzu-the-miracle-vine/.

A 1948 *Broadcasting Magazine* issue: Bernard Platt, "A Continuing Study of Major Radio Markets," *Broadcasting Magazine* (August 2, 1948), https://worldradiohistory.com/Archive -BC/BC-1948/1948-08-02-BC-Atlanta.pdf.

And a 2017 piece: Clinton Crockett Peters, "The Miracle Vine," The Awl, August 29, 2017, https://www.theawl.com/2017/08/kudzu-the-miracle-vine/.

Whether it was verses: Channing Cope, *Front Porch Farmer* (Turner E. Smith & Co., 1946), https://babel.hathitrust.org/cgi/pt?id=c00.31924003340340&view=1up&seq=9&q1=ruthie.

Cultural geographer Alderman makes the case: Derek H. Alderman, "Channing Cope and the Making of a Miracle Vine," *Geographical Review* 94, no. 2 (2004): 157–77. http://www.jstor .org/stable/30033969.

167   **It wasn't long before the media personality:** "The Press: The Kudzu Kid," *TIME.Com* (July 4, 1949), https://content.time.com/time/subscriber/article/0,33009,888578-1,00.html.

It was also in the middle: J. Farley, "Mosquitoes or Malaria? Rockefeller Campaigns in the American South and Sardinia," *Parassitologia* (August 1994), https://pubmed.ncbi.nlm.nih .gov/7898953/.

168   **By then an ardent kudzu devotee:** Kristen Hinman, "Kudzu—Japan's Wonder Vine," *History Net* (September 29, 2017), https://www.historynet.com/kudzu-japans-wonder-vine/.

That same Georgia-based initiative: Susan Howson, "Kudzu," *Quartz* (July 21, 2022), https://qz.com/emails/quartz-obsession/2089731/kudzu.

It was on the kudzu-kissed soil: Derek H. Alderman, "Channing Cope and the Making of a Miracle Vine," *Geographical Review* 94, no. 2 (2004): 157–77. http://www.jstor.org/stable /30033969.

Thanks to the next: Kristen Hinman, "Kudzu—Japan's Wonder Vine," *HistoryNet* (September 29, 2017), https://www.historynet.com/kudzu-japans-wonder-vine/.

"If it hadn't been for kudzu": Staff, "Kudzu: We've Made a Big Mistake," *Quartz* (November 16, 2021), https://qz.com/2084567/what-the-invasive-vine-kudzu-can-teach-us -about-climate-change/#:%7E:text=%5Brecording%20of%20Channing%20Cope%3A%20 %E2%80%9C,anymore%E2%80%94kudzu%20is%20king.%5D.

170   **Another of his journal articles:** Derek H. Alderman, "When an Exotic Becomes Native: Taming, Naming, and Kudzu as Regional Symbolic Capital," *Southeastern Geographer* 55, no. 1 (2015): 32–56, https://doi.org/10.1353/sgo.2015.0004.

14. ALIENS IN AMERICA

171   **"I [have] thought of kudzu as another form of COVID":** Julius Thompson, interview with author, June 2, 2022, Statham, GA.

       Killer Kudzu **is a book of metaphors:** Julius Thompson, *Killer Kudzu* (Julius Thompson, 2022).

173   **There, in the summer of 2020:** Coco Romack, "The Artist Who Transforms Galleries into Forests and Fields," *New York Times* (May 5, 2021), https://www.nytimes.com/2021/05/03/t-magazine/precious-okoyomon-artist-shed.html.

       A booklet that ran with the exhibition: Hannah Black and Precious Okoyomon, "Precious Okoyomon: Earthseed," Museum Für Moderne Kunst, 2020, https://cms.mmk.art/site/assets/files/5224/mmk_booklet_earthseed_en-1.pdf.

       In an article for the *New York Times:* Alice Walker, "Ten Years after the March on Washington," *New York Times* August 26, 1973, https://www.nytimes.com/1973/08/26/archives/staying-home-in-mississippi-ten-years-after-the-march-on-washington.html.

174   **For a 1998 study:** Kathleen S. Lowney and Joel Best, "Floral Entrepreneurs: Kudzu as Agricultural Solution and Ecological Problem," *Sociological Spectrum* 18, no. 1 (1998): 93–114, https://doi.org/10.1080/02732173.1998.9982186.

       It has even been used symbolically: Heath R. Robertson, "Cultural Kudzu: The Creep of an Invasive Culture Upon the Cherokee," *Journal of Thought* 55, no. 1/2 (2021): 57–68, https://www.proquest.com/openview/53c819e7258bf4f575f55cf08c1eb7bf/1?pq-origsite=gscholar&cbl=48333.

       On December 7, 1941: "Pearl Harbor Attack," *Encyclopaedia Britannica* (April 23, 2023).

175   **Later ruled in the Tokyo Trials:** "Tokyo War Crimes Trial," The National WWII Museum, New Orleans, n.d., https://www.nationalww2museum.org/war/topics/tokyo-war-crimes-trial.

       The federal government launched: The Children of the Camps Project, "Children of the Camps: Internment History," PBS, 1999, https://www.pbs.org/childofcamp/history/.

       where at least 110,000 people: Gwynn Guilford, "The Dangerous Economics of Racial Resentment during World War II," *Quartz* (February 13, 2018), https://qz.com/1201502/japanese-internment-camps-during-world-war-ii-are-a-lesson-in-the-scary-economics-of-racial-resentment/.

       It is indisputable: Boris Heersink and Jeffery A. Jenkins, "The Republican Party Is White and Southern. How Did That Happen?" *Washington Post* (February 7, 2020), https://www.washingtonpost.com/politics/2020/02/07/republican-party-is-white-southern-how-did-that-happen/.

       fueled by systemic racism: Li Zhou, "Why Violence against the Asian American Community Is on the Rise during the Covid-19 Pandemic," *Vox* (March 5, 2021), https://www.vox.com/identities/2020/4/21/21221007/anti-asian-racism-coronavirus-xenophobia.

       buoyed by America's conflicting power: Kevin Rudd, "How to Stop China and the US Going to War," *Guardian* (April 15, 2022), https://www.theguardian.com/world/2022/apr/07/how-to-stop-china-and-the-us-going-to-war.

       Tension has heightened: Sigal Samuel, "What Is a Wet Market? Here's Why China Is Reopening Them despite Coronavirus," *Vox* (April 15, 2020), https://www.vox.com/future-perfect/2020/4/15/21219222/coronavirus-china-ban-wet-markets-reopening.

perpetrated by a majority of white attackers: Kimmy Yam, "Safety Concerns over Anti-Asian Hate Keep Students Learning from Home," *NBC News* (June 15, 2021), https://www.nbcnews.com/news/asian-america/viral-images-show-people-color-anti-asian-perpetrators-misses-big-n1270821.

Between March 2020 and March 2022: Edwin Rios, "Hate Incidents against Asian Americans Continue to Surge, Study Finds," *Guardian* (December 15, 2022), https://www.theguardian.com/us-news/2022/jul/21/asian-americans-hate-incidents-study.

In early 2023: Emily Feng, "How a Chinese 'Spy Balloon' Prompted the U.S. to Scour the Skies," *NPR* (February 14, 2023), https://www.npr.org/2023/02/14/1156731462/china-spy-balloon-timeline-key-dates.

A 2010 paper: Anna Eskridge and Derek H. Alderman, "Alien Invaders, Plant Thugs, and the Southern Curse," *Southeastern Geographer* 50, no. 1 (2010), http://www.jstor.org/stable/26225593.

176 **"If you think about it, this discourse resonates with a lot of Americans":** Derek Alderman, interview with author, May 24, 2022, Knoxville, TN.

A place where "alien" has been used: Jake Scobey-Thal, "Illegal Alien: A Short History," *Foreign Policy* (August 27, 2014), https://foreignpolicy.com/2014/08/27/illegal-alien-a-short-history/.

It is not just wielded: "Alien," *Merriam-Webster Dictionary* (2023), https://www.merriam-webster.com/dictionary/alien.

"There's different schools of thought on the extent to which language shapes our thinking": Sarah Rich, interview with author, May 25, 2022, Atlanta, GA.

177 **A trend in modern American immigration rhetoric:** Steve Rose, "A Deadly Ideology: How the 'Great Replacement Theory' Went Mainstream," *Guardian* (June 8, 2022), https://www.theguardian.com/world/2022/jun/08/a-deadly-ideology-how-the-great-replacement-theory-went-mainstream.

This alarmingly racist: Dustin Jones, "What Is the 'Great Replacement' and How Is It Tied to the Buffalo Shooting Suspect?" *NPR* (May 16, 2022), https://www.npr.org/2022/05/16/1099034094/what-is-the-great-replacement-theory.

The theory was linked to: Alan Feuer, "How Buffalo Suspect's Racist Writings Reveal Links to Other Attacks," *New York Times* (May 16, 2022), https://www.nytimes.com/2022/05/16/us/buffalo-shooting-replacement-theory-christchurch-el-paso.html.

wounding dozens: Cindy Ramirez, "El Paso Walmart Shooting Survivor Recounts Horror, Tragedy on Anniversary," *Texas Tribune* (August 3, 2022), https://www.texastribune.org/2022/08/03/el-paso-walmart-shooting-survivor/.

On October 27, 2018: Jason Silverstein, "Pittsburgh Synagogue Shooting Suspect Identified as Robert Bowers: What We Know," *CBS News* (August 21, 2020), https://www.cbsnews.com/news/pittsburgh-synagogue-shooting-suspect-identified-as-robert-bowers-what-we-know-2018-10-27/.

On March 15, 2019: Emanuel Stoakes, "New Zealand Mosque Attack Victims Confront Gunman in Courtroom," *Washington Post* (August 24, 2020), https://www.washington

post.com/world/asia_pacific/brenton-tarrant-sentence-new-zealand-mosque-attack
-christchurch/2020/08/23/abd51832-e10c-11ea-82d8-5e55d47e90ca_story.html.

all while live-streaming it: Ryan Mac, Kellen Browning, and Sheera Frenkel, "Livestreams
of Mass Shootings: From Buffalo to New Zealand," *New York Times* (May 20, 2022), https://
www.nytimes.com/2022/05/19/technology/mass-shootings-livestream-online.html.

A manifesto produced by the killer: "The 'Great Replacement' Theory, Explained," National Immigration Forum, n.d., https://immigrationforum.org/wp-content/uploads/2021
/12/Replacement-Theory-Explainer-1122.pdf.

"Anti-alien" concepts like these: Amanda Seitz, "White Supremacists Are Riling up Thousands on Social Media," *PBS NewsHour* (June 10, 2022), https://www.pbs.org/newshour
/politics/white-supremacists-are-riling-up-thousands-on-social-media.

a 2022 investigation: Nicholas Confessore, "How Tucker Carlson Stoked White Fear to
Conquer Cable," *New York Times* (May 4, 2022), https://www.nytimes.com/2022/04/30/us
/tucker-carlson-gop-republican-party.html.

178   **immigration was being purposefully exploited:** Steve Rose, "A Deadly Ideology: How
the 'Great Replacement Theory' Went Mainstream," *Guardian* (June 8, 2022), https://www
.theguardian.com/world/2022/jun/08/a-deadly-ideology-how-the-great-replacement
-theory-went-mainstream.

Exclusionary restrictions: "How U.S. Immigration Laws and Rules Have Changed
through History," Pew Research Center, May 30, 2020, https://www.pewresearch.org/fact
-tank/2015/09/30/how-u-s-immigration-laws-and-rules-have-changed-through-history/.

In her memoir: Maria Hinojosa, *Once I Was You: A Memoir* (Simon and Schuster, 2020).

Twenty years later: "Chinese Exclusion Act (1882)," National Archives, January 17, 2023,
https://www.archives.gov/milestone-documents/chinese-exclusion-act.

179   **It would remain the baseline until the 1920s:** "The Immigration Act of 1924 (The Johnson-
Reed Act)," Office of the Historian, Foreign Service Institute, United States Department of
State, n.d., https://history.state.gov/milestones/1921-1936/immigration-act.

In 1943, when the US needed: Jane H. Hong, *Opening the Gates to Asia: A Transpacific History of How America Repealed Asian Exclusion* (University of North Carolina Press, 2019).

Lest we forget: Mark M. Davis, Matthew Ming-Tak Chew, Richard J. Hobbs, Ariel E. Lugo,
John J. Ewel, Geerat J. Vermeij, James H. Brown, et al., "Don't Judge Species on Their Origins," *Nature* 474, no. 7350 (2011): 153–54, https://doi.org/10.1038/474153a.

began applying this dichotomy: John Stevens Henslow, 1835, *The Principles of Descriptive
and Physiological Botany* (Longman, 2008).

Scientists first proposed: Christopher Preston, David Pearman, and Allan Hall, "Archaeo-
phytes in Britain," *Botanical Journal of the Linnean Society* 145, no. 3: (2004): 257–94, https://
doi.org/10.1111/j.1095-8339.2004.00284.x.

Nearly two centuries later: The Invasive Species Specialist Group, "Environmental Impact
Classification for Alien Taxa (EICAT)," Global Invasive Species Database, 1996, http://www
.iucngisd.org/gisd/about_eicat.php.

Her 2022 paper: Katrina Maggiulli, "Teaching Invasive Species Ethically: Using Comics to
Resist Metaphors of Moral Wrongdoing & Build Literacy in Environmental Ethics," *Environmental Education Research* 28, no. 9 (2022), https://doi.org/10.1080/13504622.2022.2085247.

180 **"It's a huge gap":** Katrina Maggiulli, interview with author, June 27, 2022, Eugene, OR.

The emblematic species is: Forest and Rangeland Ecosystem Science Center, "Threat of Invasive Barred Owls to Northern Spotted Owls and Their Habitats," US Geological Survey, November 14, 2017, https://www.usgs.gov/centers/forest-and-rangeland-ecosystem -science-center/science/threat-invasive-barred-owls-northern#overview.

181 **Despite the barred owls' ability:** Allison Frost, "Saving Endangered Spotted Owls Means Killing Some Barred Owls," *Oregon Public Broadcasting* (July 30, 2021), https://www.opb.org /article/2021/07/30/saving-endangered-spotted-owls-means-killing-some-barred-owls/.

The agency released: Forest and Rangeland Ecosystem Science Center, "Effects of Experimental Removal of Barred Owls on Population Demography of Northern Spotted Owls in the Pacific Northwest," US Geological Survey, January 8, 2018, https://www.usgs.gov /centers/forest-and-rangeland-ecosystem-science-center/science/effects-experimental -removal-barred.

A US federal appeals court: "Friends of Animals v. US Fish and Wildlife Service," United States Court of Appeals for the Ninth Circuit, November 17, 2021, https://cases.justia.com /federal/appellate-courts/ca9/21-35062/21-35062-2022-03-04.pdf?ts=1646416989.

183 **Journalist Jenny Morber:** Jenny Morber, "Should Plants and Animals That Relocate Because of Climate Change Be Considered Invasive?" *Ensia* (July 30, 2020), https://ensia.com /features/climate-change-nonnative-invasive-species/.

That research supports the idea: "Environmental Impact Classification for Alien Taxa (EICAT)," International Union for Conservation of Nature and Natural Resources, n.d., https://www.iucn.org/resources/conservation-tool/environmental-impact-classification -alien-taxa#:~:text=The%20Environmental%20Impact%20Classification%20for,living %20outside%20their%20natural%20orange.

Rather than a universal standard: Piper D. Wallingford, Toni Lyn Morelli, Jenica M. Allen, Evelyn M. Beaury, Dana M. Blumenthal, Bethany A. Bradley, Jeffrey S. Dukes, et al., "Adjusting the Lens of Invasion Biology to Focus on the Impacts of Climate-Driven Range Shifts," *Nature Climate Change* 10, no. 5 (2020): 398–405, https://doi.org/10.1038/s41558-020 -0768-2.

And a 2018 *Smithsonian* article: Amy Crawford, "Why We Should Rethink How We Talk About 'Alien' Species," *Smithsonian Magazine* (January 9, 2018), https://www.smithsonian mag.com/science-nature/why-scientists-are-starting-rethink-how-they-talk-about-alien -species-180967761/.

"Don't Judge Species on Their Origins": Mark M. Davis, Matthew Ming-Tak Chew, Richard J. Hobbs, Ariel E. Lugo, John J. Ewel, Geerat J. Vermeij, James H. Brown, et al., "Don't Judge Species on Their Origins," *Nature* 474, no. 7350 (2011): 153–54, https://doi.org/10.1038 /474153a.

184 **Following the failed crusade:** "Southwestern Willow Flycatcher (Empidonax Traillii Extimus)," Environmental Conservation Online System, US Fish and Wildlife Service, n.d., https://ecos.fws.gov/ecp/species/6749.

Advocating for official changes: The Nature Conservancy, "The movement to rename species," August 25, 2023. https://www.nature.org/en-us/magazine/magazine-articles/new -common-names/.

A surge of attention: Jessie Molloy, "The Real Reason Asian Carp Is Being Renamed," Tasting Table, February 2, 2023. https://www.tastingtable.com/905575/the-real-reason-asian-carp-is-being-renamed/.

On smaller scales: Rutgers, The State University of New Jersey, Michele Bakacs, and William Errickson. "FS1353: Invasive Plants and Native Alternatives for Landscapes (Rutgers NJAES)." New Jersey Agricultural Experiment Station, August 2023. https://njaes.rutgers.edu/fs1353/.

The word is even incorporated into: Rod Randall, "Pueraria Lobata (Fabaceae)," Global Compendium of Weeds (GCW), Hawaiian Ecosystems at Risk (HEAR), September 13, 2007, http://www.hear.org/gcw/species/pueraria_lobata/.

According to the GCW: R.P. (Rod) Randall, "A Global Compendium of Weeds," Hawaiian Ecosystems at Risk (HEAR), n.d., http://www.hear.org/gcw/pdfs/gcw_intro.pdf.

Fourteen percent of all weeds: Gisella S. Cruz-Garcia and Lisa L. Price, "Weeds as Important Vegetables for Farmers," *Acta Societatis Botanicorum Poloniae* (December 2012), https://doi.org/10.5586/asbp.2012.047.

They're also costly: "Summary for Policymakers of the Thematic Assessment Report on Invasive Alien Species and Their Control of the Intergovernmental Science-Policy Platform on Biodiversity and Ecosystem Services." *IPBES,* September 4, 2023. https://www.ipbes.net/IASmediarelease#:~:text=Approved%20on%20Saturday%20in%20Bonn,%24423%20billion%20annually%20in%202019%2C.

From rats to goats: Patrick Greenfield, "Driving out Invasive Species on Islands Has High Success Rate and Big Benefits—Study," *Guardian* (August 10, 2022), https://www.theguardian.com/environment/2022/aug/10/driving-out-invasive-species-on-islands-high-success-rate-big-benefits-aoe.

The ecological impacts: Dena R. Spatz, Nick D. Holmes, David William Will, Stella Hein, Zachary T. Carter, Rachel M. Fewster, Bradford S. Keitt, et al., "The Global Contribution of Invasive Vertebrate Eradication as a Key Island Restoration Tool," *Scientific Reports* 12, no. 1 (2022), https://doi.org/10.1038/s41598-022-14982-5.

The sometimes 200-pound snakes: "Burmese Python (Python Bivittatus)—Species Profile," USGS Nonindigenous Aquatic Species Database, June 27, 2022, https://nas.er.usgs.gov/queries/FactSheet.aspx?speciesID=2552.

First spotted in 1979: Radha Krueger, "Burmese Python," Rare, Beautiful & Fascinating (collection), Florida Museum, University of Florida, August 28, 2017, https://www.floridamuseum.ufl.edu/100years/burmese-python/.

A 2012 study: Michael E. Dorcas, John D. Willson, Robert A. Reed, Ray W. Snow, Michael R. Rochford, Melissa A. Miller, Walter E. Meshaka, et al., "Severe Mammal Declines Coincide with Proliferation of Invasive Burmese Pythons in Everglades National Park," *Proceedings of the National Academy of Sciences of the United States of America* 109, no. 7 (2012): 2418–22, https://doi.org/10.1073/pnas.1115226109.

The evidence of its impact: "How Have Invasive Pythons Impacted Florida Ecosystems?" US Geological Survey, April 5, 2019, https://www.usgs.gov/faqs/how-have-invasive-pythons-impacted-florida-ecosystems.

185

186 **Except these snakes:** Sarah Watts, "Jeffrey Dahmer, Albert Fish and More: Why Do Some Serial Killers Turn into Cannibals?" A+E Networks, September 17, 2018, https://www .aetv.com/real-crime/cannibal-serial-killers-jeffrey-dahmer-albert-fish-andrei-chikatilo -psychology-of-cannibalism.

One example: USGS Nonindigenous Aquatic Species Database. "Burmese Python (Python Bivittatus) - Species Profile," June 27, 2022. https://nas.er.usgs.gov/queries/FactSheet .aspx?speciesID=2552.

A 2020 essay penned by Jenny Liou: Jenny Liou, "Am I an Invasive Species?" *High Country News* (July 9, 2020), https://www.hcn.org/issues/52.8/north-race-and-racism-am-i-an -invasive-species.

15. AM I INVASIVE?

191 **Once, America represented a gateway to freedom:** "U.S. Immigration Before 1965," *History* (October 29, 2009), updated September 10, 2021, https://www.history.com/topics /immigration/u-s-immigration-before-1965.

It was white Europeans: Lois Thielen, "For the Founding Fathers, Only White Men Deserved Rights," *SCTimes* (July 1, 2019), https://eu.sctimes.com/story/opinion/2019/07/01 /founding-fathers-only-white-men-deserved-rights/1620597001/.

We see it in the disproportionate: Lynne Peeples, "What the Data Say about Police Brutality and Racial Bias—and Which Reforms Might Work," *Nature* 583, no. 7814 (2020): 22–24, https://doi.org/10.1038/d41586-020-01846-z.

entrenched in racism and prejudice: Robert S. McElvaine, "Racism, Policing, Politics and Violence: How America in 2022 Was Shaped by 1964," *Salon* (July 21, 2022), https://www .salon.com/2022/07/21/racism-policing-and-violence-how-america-in-2022-was-shaped -by-1964/.

We see it in the American incarceration system: Ashley Nellis, "The Color of Justice: Racial and Ethnic Disparity in State Prisons," *The Sentencing Project* (December 2022), https:// www.sentencingproject.org/publications/color-of-justice-racial-and-ethnic-disparity -in-state-prisons/.

192 **We see it reflected in the makeup of our legislators:** Anagha Srikanth, "New Study Finds White Male Minority Rule Dominates US," *The Hill* (May 26, 2021), https://thehill.com /changing-america/respect/diversity-inclusion/555503-new-study-finds-white-male -minority-rule/.

193 **"It's a marginalized plant, a plant that loves the edges,'":** Justin Holt, interview with author, January 13, 2021, Asheville, NC.

194 **According to the IPBES:** Patrick Greenfield and Phoebe Weston, "The Five Biggest Threats to Our Natural World . . . and How We Can Stop Them," *Guardian* (December 4, 2022), https:// www.theguardian.com/environment/2021/oct/14/five-biggest-threats-natural-world-how -we-can-stop-them-aoe.

A NOTE FROM THE AUTHOR

197   **Flames flickered at:** Robyn Dixon, "Siberia's Wildfires Are Bigger Than All the World's Other Blazes Combined," *Washington Post,* August 12, 2021, https://www.washingtonpost.com/world/2021/08/11/siberia-fires-russia-climate/.

         Frequent droughts: Laura Paddison, "Catastrophic Drought That's Pushed Millions into Crisis Made 100 Times More Likely by Climate Change, Analysis Finds." CNN, April 27, 2023, https://www.cnn.com/2023/04/27/africa/drought-horn-of-africa-climate-change-intl.

         Across the Global South: UNHCR US. "Climate Change and Disaster Displacement | UNHCR US." UNHCR US, n.d., https://www.unhcr.org/us/what-we-do/how-we-work/environment-disasters-and-climate-change/climate-change-and-disaster.

         And those burdens: Rachel Morello-Frosch and Osagie K Obasogie, "The Climate Gap and the Color Line—Racial Health Inequities and Climate Change," *New England Journal of Medicine* 388, no. 10 (March 9, 2023): 943–49, https://doi.org/10.1056/nejmsb2213250.

198   **There is still a window of opportunity:** NOAA Climate.gov, "Can We Slow or Even Reverse Global Warming?," October 12, 2022, https://www.climate.gov/news-features/climate-qa/can-we-slow-or-even-reverse-global-warming.

Index